AFOQT Math Prep

The Ultimate Step-by-Step Guide Plus
Two Full-Length AFOQT Practice Tests

SCAN ME

Michael Smith
www.mathnotion.com

AFOQT Math Prep

Published By: The Math Notion

Web: www.mathnotion.com

Email: info@mathnotion.com

ISBN: 978-1-63620-189-4

The Math Notion

Michael Smith has been a math instructor for over a decade now. He launched the Math Notion. Since 2006, we have devoted our time to both teaching and developing exceptional math learning materials. As a test prep company, we have worked with thousands of students. We have used the feedback of our students to develop a unique study program that can be used by students to drastically improve their math scores fast and effectively. We have more than a thousand Math learning books including:

- **ASVAB Math Prep**
- **GED Math Prep**
- **HiSET Math Prep**
- **TABE Math Prep**
- **TASC Math Prep**
- **many Math Education Workbooks, Study Guides, Practice and Exercise Books**

As an experienced Math test preparation company, we have helped many students raise their standardized test scores—and attend the colleges of their dreams: We tutor online and in person, we teach students in large groups, and we provide training materials and textbooks through our website and through Amazon.

You can contact us via email at:

info@mathnotion.com

How to Achieve a Perfect Score on the AFOQT Math Test?

AFOQT Math Prep covers all mathematics topics that will be key to succeeding on the AFOQT math test. The step-by-step guide and hundreds of examples in this book can help you hone your math skills, boost your confidence, and be well prepared for the AFOQT test.

This new AFOQT self-teaching prep book offers extensive preparation and brush-up in math for those test-takers who plan to take the AFOQT test. Two AFOQT math practice exams with detailed answers reflect questions and question types found on the actual test.

Updated the new AFOQT prep for 2020, 2021, and beyond by top test prep experts. Inside the AFOQT math preparation, You'll Find:

- Comprehensive AFOQT math review of key concepts

- Content 100% aligned with the latest AFOQT math exam

- Over 2,000 practice questions to help you master each AFOQT math topic

- Lots of examples with step-by-step solutions to illustrate all the AFOQT question types

- Each lesson is short, concise, and to the point.

- Focus on the most challenging part of the AFOQT math test!

- 2 full-length practice exams (featuring new question types) with detailed answers

- And much more than we can fit in this space…

After completing this math preparation book, you will discover your strengths and weaknesses, a strong foundation, and gain confidence to be successful on the AFOQT test. We have helped hundreds of thousands of people pass the AFOQT test and achieve their education and career goals. Get math prep for the AFOQT review that you need to ace your exam.

It is an excellent investment in your future!

www.MathNotion.com

... So Much More Online!

✓ FREE Math Lessons

✓ More Math Learning Books!

✓ Mathematics Worksheets

✓ Online Math Tutors

✓ For a PDF Version of This Book

SCAN ME

Please Visit www.mathnotion.com

Contents

Chapter 1 : Whole Numbers, Real Numbers, and Integers

Topics that you'll learn in this chapter:

> ➢ Rounding and Estimates

> ➢ Addition, Subtraction, Multiplication and Division Whole Number and
>
> Integers

> ➢ Arrange and ordering Integers and Numbers

> ➢ Comparing Integers, Order of Operations

> ➢ Mixed Integer Computations

> ➢ Integers and Absolute Value

"If people do not believe that mathematics is simple, it is only because they do not realize how complicated life is." — John von Neumann

Name: ....

Rounding

Rounding is replacing a number up or down to the closest number or the closest hundred, etc.

- ✓ First, you have to know the place value you'll round to.
- ✓ Second, you have to find the digit to the right of the place value you're rounding to. If it is 5 or greater, add 1 to the place value you're rounding to and put zero for all digits on its right side. If the digit to the right of the place value is smaller than 5 then keep the place value and put zero for all digits to the right.

EXAMPLE:

Round 64 to the closest ten.

The place value of ten is 6. The digit on the right side is 4 (which is smaller than 5). Now keep 6 and put zero for the digit on the right side. Now our answer is 60. 64 is rounded to the closest ten is 60, because 64 is closer to 60 than to 70.

PRACTICES:

Round each number to the underlined place value.

1) 88	2) 8.15
3) 4,315	4) 565
5) 1.331	6) 14.23
7) 2.429	8) 4.313
9) 2.997	10) 7.38

Score: ....

	Answer Key	
1) 90	2) 8.0	
3) 4,000	4) 570	
5) 1.3	6) 14.2	
7) 2.0	8) 4.31	
9) 3.0	10) 7.0	

Name: ....

Estimates

Estimating is a math policy used for approximating a number. To estimate *means* to make an irregular guess or calculation. To round means to make easier a known number by scaling it a little bit up or down.

- ✓ To estimate a math problem, round the numbers.
- ✓ For 2-digit numbers, you can usually round to the nearest tens, for 3-digit numbers, round to nearest hundreds, etc.
- ✓ Find the answer.

EXAMPLE:

Estimate the sum by rounding every number to the closest hundred. **153 + 426 =?**

153 is rounded to the closest hundred which is 200. Now 426 is rounded to the closest hundred which is 400.

Then: 200 + 400 = 600

PRACTICES:

Estimate the sum by rounding each added to the nearest ten.

1) 17 + 18	2) 94 + 81
3) 203 + 56	4) 55 + 33
5) 96 + 49	6) 99 + 324
7) 823 + 488	8) 466 + 276
9) 5,112 + 5,792	10) 1,245 + 2,459

Score: ....

Answer Key	
1) 40	2) 200
3) 260	4) 90
5) 150	6) 400
7) 1,300	8) 800
9) 11,000	10) 3,000

Name: ....

Whole Number Addition and Subtraction

- ✓ Arrange the numbers in line.
- ✓ Start with the unit place. (Ones place)
- ✓ Regroup if needed.
- ✓ Add or subtract the tens place.
- ✓ Continue with further digits.

EXAMPLE:

Find the sum. **285 + 145 =?**

First line up the numbers: $\frac{285}{+145}$ → Start with the unit place. (ones place) $5 + 5 = 10$,

Write 0 for ones place and keep 1, $\overset{1}{\underset{0}{\frac{285}{+145}}}$, Add the tens place and the digit 1 we kept:

$1 + 8 + 4 = 13$, Write 3 and keep 1, $\overset{1\,1}{\underset{30}{\frac{285}{+145}}}$

Continue with further digits → $1 + 2 + 1 = 4$ → $\overset{1\,1}{\underset{430}{\frac{285}{+145}}}$

Find the difference. **976 − 453 = ?**

First line up the numbers: $\frac{976}{-453}$, → Start with the unit place. $6 - 3 = 3$, $\underset{3}{\frac{976}{-453}}$,

Subtract the tens place. $7 - 5 = 2$, $\underset{23}{\frac{976}{-453}}$, Continue with further digits → $9 - 4 = 5$,

$\underset{523}{\frac{976}{-453}}$

PRACTICES:

Find the missing number.

1) 540 − .... = 100	2) 800 − .... = 220
3) .... − 2,650 = 6,700	4) 85,000 − 42,000 = ...
5) 1,280− .... = 420	6) 5,000 + 8,450 = ...
7) .... − 3,870 = 9,630	8) 12,310 − .... = 8,540

Solve.

9) A school had 708 students last year. If all last year students and 218 new students have registered for this year, how many students will there be in total?

10) Lisa had $856 dollars in her saving account. She gave $295 dollars to her brother, Tom. How much money does she have left?

Score: ....

Answer Key	
1) 440	2) 580
3) 9,350	4) 43,000
5) 860	6) 13,450
7) 13,500	8) 3,770
9) 926	10) 561

Name: ....

Whole Number Multiplication

✓ First you have to learn the times tables! To solve multiplication problems quick, you need to learn the times table. For example, 3 times 8 is 24 or 8 times 7 is 56.

✓ For multiplication, line up the numbers that you are multiplying.

✓ Start with the ones place and regroup if needed.

✓ Continue with further digits.

EXAMPLE:

Solve. $500 \times 30 = ?$

Line up the numbers: $\begin{array}{r} 500 \\ \times 30 \\ \hline \end{array}$, start with the ones place $\rightarrow 0 \times 500 = 0$, $\begin{array}{r} 500 \\ \times 30 \\ \hline 0 \end{array}$, Continue

with further digit which is 3. $\rightarrow 3 \times 500 = 1,500$, $\begin{array}{r} 500 \\ \times 30 \\ \hline 15,000 \end{array}$

PRACTICES:

Multiply the Number.

1) $120 \times 6 =$ _____	2) $160 \times 30 =$ _____
3) $600 \times 30 =$ _____	4) $420 \times 20 =$ _____
5) $250 \times 40 =$ _____	6) $600 \times 40 =$ _____
7) $215 \times 70 =$ _____	8) $540 \times 11 =$ _____
9) $121 \times 10 =$ _____	10) $254 \times 16 =$ _____

Score: ....

Answer Key	
1) 720	2) 4,800
3) 18,000	4) 8,400
5) 10,000	6) 24,000
7) 15,050	8) 5,940
9) 1,210	10) 4,064

Name: ....

Whole Number Division

- ✓ Division: A typical division problem: Dividend ÷ Divisor = Quotient
- ✓ In division, we want to find how many times a divisor is contained in a dividend. The result we obtain in a division problem is called quotient.
- ✓ First, the problem is written in division format. (Dividend is inside; divisor is outside)

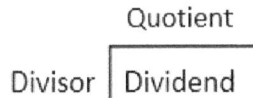

$$\text{Divisor} \overline{\big)\ \text{Dividend}}^{\text{Quotient}}$$

EXAMPLE:

Solve. $234 \div 4 = ?$

First, write the problem in division format. $4\overline{\big)\ 234}$

Start from left digit of the dividend. 4 won't divide 2.

So, we have to choose another digit of the dividend. It is 3.

Now, we will find how many times 4 goes into 23 and the answer is 5.

Write 5 above the dividend part. 4 times 5 is 20.

$$4\overline{\big)\ 234}^{\ 5}$$

Write 20 below 23 and subtract. We get the answer 3.

Now take down the next digit which is 4 and find how many times 4 goes into 34?

The answer is 8. Write 8 above dividend.

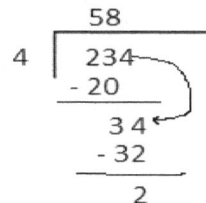

This is last step since there is no further digit left.

of the dividend to bring down.

The final answer is 58 and we have the remainder 2.

$$
\begin{array}{r}
58 \\
4\overline{\big)\ 234} \\
-20 \\
\hline
34 \\
-32 \\
\hline
2
\end{array}
$$

PRACTICES:

Divide the Number.

1) $450 \div 5 =$ _____	2) $320 \div 8 =$ _____
3) $125 \div 25 =$ _____	4) $720 \div 12 =$ _____
5) $588 \div 14 =$ _____	6) $299 \div 13 =$ _____
7) $869 \div 11 =$ _____	8) $801 \div 9 =$ _____
9) $493 \div 17 =$ _____	10) $600 \div 24 =$ _____

Score: ....

Answer Key	
1) 90	2) 40
3) 5	4) 60
5) 42	6) 23
7) 79	8) 89
9) 29	10) 25

Name: ....

Adding and Subtracting Integers

- ✓ Integers include zero, positive natural numbers, and the negative of the natural numbers. $\{... , -3, -2, -1, 0, 1, 2, 3, ...\}$
- ✓ Add a positive integer by putting it to the right on the number line.
- ✓ Add a negative integer by putting it to the left on the number line.
- ✓ Subtract an integer by adding its opposite.

EXAMPLE:

Solve. $(-8) - (-5) =$

We keep the first number and change the sign of the second number to its opposite.

(Change subtraction into addition. Then: $(-8) + 5 = -3$

Solve. $10 + (4 - 8) =$

First subtract the numbers in brackets, $4 - 8 = -4$

Then: $10 + (-4) = \rightarrow$ changes addition into subtraction: $10 - 4 = 6$

PRACTICES:

Find the sum and difference.

1) $8 + (-11)$	2) $(-13) + 25$
3) $(55) - (21)$	4) $(4) - (-5) - (-3)$
5) $2 + (-11) + (-30) + (9)$	6) $(-5) + (-10) + (7-19)$
7) $(-20) - (-44)$	8) $(-9) - 13 + 20$
9) $(50) - (-5) + (-25)$	10) $24 + 16 + (-13)$

Score: ....

Answer Key	
1) −3	2) 12
3) 34	4) 12
5) −30	6) −27
7) 24	8) −2
9) 30	10) 27

Name: ....

Multiplying and Dividing Integers

- ✓ (positive) × (positive) = positive
- ✓ (positive) ÷ (positive) = positive
- ✓ (negative) × (negative) = positive
- ✓ (negative) ÷ (negative) = positive
- ✓ (negative) × (positive) = negative
- ✓ (negative) ÷ (positive) = negative
- ✓ (positive) × (negative) = negative
- ✓ (positive) ÷ (negative) = negative

÷ ╱ ×	+	−
+	+	−
−	−	+

EXAMPLE:

$(+5) \times (+3) = 5 + 5 + 5 = 15$

The basic idea of multiplication is recurrent addition. Example: $5 \times 3 = 5 + 5 + 5 = 15$

We know that division is the inverse operation of multiplication. So, $15 \div 3 = 5$ because $5 \times 3 = 15$ In words, this expression says that 15 may be divided into 3 groups of 5 every because adding five thrice gives 15.

Divide (-91) by (-7)?

Examples on division of integers on different kinds of problems on integers are mentioned here step by step. $(-91) \div (-7) = 13$

PRACTICES:

Find each product and each quotient.

1) $(-8) \times (-5)$	2) $72 \div 9$
3) $4 \times (-5) \times (-6)$	4) $(-95) \div (-5)$
5) $32 \times (-4)$	6) $(-99) \div (-11)$
7) $(-12) \times (-4)$	8) $(-123) \div 1$
9) $(-4) \times (-3) \times 5$	10) $(-0) \div 15$

Score: ....

Answer Key	
1) 40	2) 8
3) 120	4) 19
5) −128	6) 9
7) 48	8) −123
9) 60	10) 0

Name: ....

Arrange, Order, and Comparing Integers

✓ When we use a number line, numbers are increased when you go to the right.
✓ To compare numbers, you can use number line! As you go from left to right on the number line, you will find a greater number!
✓ Order integers from smallest to greatest.

EXAMPLE:

Order integers from to greatest.

$$(-11, -13, 7, -2, 12)$$

To compare numbers, you can use number line! When you see from left to right on the number line, you find a greater number!

$$-13 < -11 < -2 < 7 < 12$$

PRACTICES:

Order each set of integers from least to greatest.	Order each set of integers from greatest to least
1) $2, 6, -15, -11, 1$	2) $1, 17, 6, 8, 65, 2$
3) $9, -8, 3, -2, 11$	4) $-12, 6, -7, 2, -11$
5) $36, -12, 5, 1, -2$	6) $-14, 17, 7, 37, 9$
7) $31, 18, 0, -54, 9, -5$	8) $-54, 0, 14, 19, 15$
9) $-15, -25, -37, 7, 0, 9$	10) $12, 7, -1, -11, 9, -3$

Score: ....

Answer Key

1) − 15, − 11, 1, 2, 6	2) 65, 17, 8, 6, 2, 1
3) − 8, − 2, 3, 9 ,11	4) 6, 2, − 7, − 11, − 12
5) − 12, − 2, 1, 5, 36	6) 37, 17, 9, 7, −14
7) −54, −5, 0, 9, 18, 31	8) 19, 15, 14, 0, −54
9) − 37, − 25, −15, 0, 7, 9	10) 12, 9, 7, −1, −3, −11

Name: ....

Compare Integer

✓ If you want to compare numbers, you can use a number line! As you move from left to right on the number line, you will find a greater number!

EXAMPLE:

-5 _____ -1

When we compare two integers, we use the symbols $<$ and $>$.

$-5 < -1$ means that -5 is less than -1

PRACTICES:

Compare. Use >, =, <

1) 4 _____ 3	2) -22 ____ -11
3) 0 _____ -31	4) -41 _____ -12
5) -64 _____ 64	6) -142 _____ -148
7) 68 _____ 100	8) $(-15) \times 6$ _____ $5 \times (-18)$
9) 16 _____ $-(-16)$	10) 405 _____ -405

Score: ....

Answer Key	
1) >	2) <
3) >	4) <
5) <	6) >
7) <	8) =
9) =	10) >

Name: ....

Order of Operations

When you find more than one math operation, use PEMDAS:
- ✓ Parentheses
- ✓ Exponents
- ✓ Multiplication and Division (from left to right)
- ✓ Addition and Subtraction (from left to right)

EXAMPLE:

Solve. $(11 \times 5) - (12 - 25) =$

First you have to simplify inside parentheses: $(11 \times 5) - (12 - 25) = (55) - (-13) =$

Then: $55 + 13 = 68$

PRACTICES:

Evaluate each expression.

1) $24 - (8 \times 6)$	2) $5 \times 6 - (\frac{15}{11 - (-4)})$
3) $12 - (6 \times (-3))$	4) $(6 \times 7) + (-7)$
5) $(\frac{(-1)+4}{(-1)+(-2)}) \times (-9)$	6) $\frac{30}{2(9-(-1))-10}$
7) $58 - (6 \times 9)$	8) $13 + (4 \times 2)$
9) $((-3) + 15) \div (-3)$	10) $[(-8 \div 2) \div (2 - 4))$

Score: ....

Answer Key	
1) −24	2) 29
3) 30	4) 35
5) 9	6) 3
7) 4	8) 21
9) −4	10) 2

Name: ....

Integers and Absolute Value

✓ To find a definite value of a number, simply find its distance from 0 on number line! For example, the distance of 13 and −13 from zero on number line is 13!

EXAMPLE:

Solve. $\frac{|-18|}{9} \times |5 - 8| =$

First find $|-18|$, →the definite value of -18 is 18, then: $|-18| = 18$

$\frac{18}{9} \times |5 - 8| =$

Next, we solve $|5 - 8|$, → $|5 - 8| = |-3|$, the definite value of -3 is 3. $|-3| = 3$

Then: $\frac{18}{9} \times 3 = 2 \times 3 = 6$

PRACTICES:

Write absolute value of each number.

1) 62	2) − 32
3) − 11	4) 5

Evaluate.

5) $|-12| - |3| + 2$	6) $19 + |-5 - 14| - |2|$
7) $|-11| + |-9|$	8) $|91| - |-18| - 18$
9) $|-10 + 4| \times \frac{|-7 \times 5|}{7}$	10) $\frac{|-16 \times 3|}{2} \times |-12|$

Score: ....

Answer Key

1) 62	2) 32
3) 11	4) 5
5) 11	6) 36
7) 20	8) 55
9) 30	10) 288

Chapter 2 : Fractions and Decimals

Topics that you'll learn in this chapter:

- ➤ Simplifying Fractions

- ➤ Adding and Subtracting Fractions, Mixed Numbers and Decimals

- ➤ Multiplying and Dividing Fractions, Mixed Numbers and Decimals

- ➤ Comparing and Rounding Decimals

- ➤ Converting Between Fractions, Decimals and Mixed Numbers

- ➤ Factoring Numbers, Greatest Common Factor, and Least Common Multiple

- ➤ Divisibility Rules

"A Man is like a fraction whose numerator is what he is and whose denominator is what he thinks of himself. The larger the denominator, the smaller the fraction." −Tolstoy

Name: ....

Simplifying Fractions

- ✓ Regularly divide both the top and bottom of the fraction by $2, 3, 5, 7, ...$ etc.
- ✓ Continue until you can't go any further.

EXAMPLE:

Simplify $\frac{12}{20}$.

To simplify $\frac{12}{20}$, you have to find a number that both 12 and 20 are divisible by. Both are divisible by 4. Then: $\frac{12}{20} = \frac{12 \div 4}{20 \div 4} = \frac{3}{5}$

PRACTICES:

Simplify the fractions.

1) $\frac{44}{64}$	2) $\frac{12}{26}$
3) $\frac{15}{25}$	4) $\frac{30}{45}$
5) $\frac{18}{27}$	6) $1\frac{62}{124}$
7) $4\frac{12}{66}$	8) $1\frac{55}{70}$
9) $\frac{54}{60}$	10) $7\frac{68}{136}$

Score: ....

Answer Key	
1) $\frac{11}{16}$	2) $\frac{6}{13}$
3) $\frac{3}{5}$	4) $\frac{2}{3}$
5) $\frac{2}{3}$	6) $1\frac{1}{2}$
7) $4\frac{2}{11}$	8) $1\frac{11}{14}$
9) $\frac{9}{10}$	10) $7\frac{1}{2}$

Name: ....

Factoring Numbers

✓ To break the numbers into their prime factors is called factoring.

✓ First few prime numbers are $2, 3, 5, 7, 11, 13, 17, 19$

EXAMPLE:

List all positive factors of 12.

Write the upside-down division:

The second column is the answer.

Then: $12 = 2 \times 2 \times 3$ or $12 = 2^2 \times 3$

```
12 | 2
 6 | 2
 3 | 3
 1 |
```

PRACTICES:

List all positive factors of each number.	List the prime factorization for each number.
1) 90	2) 40
3) 49	4) 105
5) 50	6) 42
7) 34	8) 78
9) 96	10) 165

Score: ....

Answer Key	
1) $1, 2, 3, 5, 6, 9, 10, 15, 18, 30, 45, 90$	2) $2 \times 2 \times 2 \times 5$
3) $1, 7, 49$	4) $3 \times 5 \times 7$
5) $1, 2, 5, 10, 25, 50$	6) $2 \times 3 \times 7$
7) $1, 2, 17, 34$	8) $2 \times 3 \times 13$
9) $1, 2, 3, 4, 6, 8, 12, 16, 24, 32, 48, 96$	10) $3 \times 5 \times 11$

Name: ....

Greatest Common Factor (GCF)

✓ List the prime factors of each number.
✓ Then multiply common prime factors.
✓ If there are no common prime factors, then our GCF is 1.

EXAMPLE:

Find the GCF for **10** and **15**.

The factors of 10 are: $\{1, 2, 5, 10\}$

The factors of 15 are: $\{1, 3, 5, 15\}$

There is 5 in common,

Then the greatest common factor is: 5

PRACTICES:

Find the GCF for each number pair.

1) 12, 25	2) 72, 84
3) 24, 36	4) 30, 45
5) 9, 36	6) 63, 42
7) 27, 12	8) 125, 50
9) 54, 39	10) 36, 52

Score: ....

	Answer Key	
1) 1		2) 12
3) 12		4) 15
5) 9		6) 21
7) 3		8) 25
9) 3		10) 4

Name: ....

Least Common Multiple (LCM)

- ✓ The smallest multiple that 2 or more numbers have in common is called least common multiple of that number. How to find LCM:
- ✓ First find the list of the prime factors of each number.
- ✓ Then multiply the common prime factors and uncommon prime factors of the numbers (each common prime factor is used only for once)

EXAMPLE:

Find the LCM for **18** and **12**.

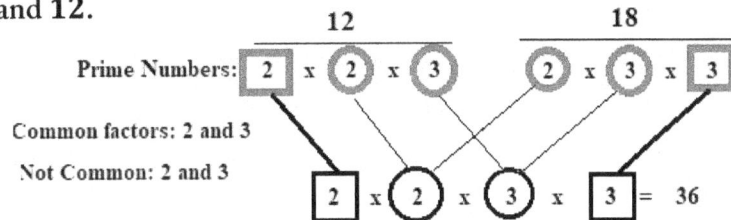

Prime Numbers:

Common factors: 2 and 3

Not Common: 2 and 3

PRACTICES:

Find the LCM for each number pair.

1) 12, 9	2) 40, 20
3) 15, 30	4) 84, 60
5) 60, 40	6) 52, 78
7) 14, 28	8) 14, 7, 42
9) 24, 32	10) 72, 66, 24

Score: ....

Answer Key	
1) 36	2) 40
3) 30	4) 420
5) 120	6) 156
7) 28	8) 42
9) 96	10) 792

Name: ....

Divisibility Rules

If a number can be divided by other numbers, it is referred as divisibility. The number is divisible:

- ✓ by 2 if the number is found even.
- ✓ by 3 if the sum of the digits is found to be divisible by 3.
- ✓ by 9 if the sum of the digits is found to be divisible by 9.
- ✓ by 4, if the last 2 digits of a number are found to be divisible by 4.
- ✓ by 6, if it is found to be divisible by 2 and 3.
- ✓ by 8, if it is found to be divisible by 2 and 4.
- ✓ by 5 if the last digit is found 0 or 5.
- ✓ by 10 if the last digit is 0.

EXAMPLE:

What is the factor of 240?

2, because the number is even   3, because $(2 + 4 + 0 = 6, 6 \div 3 = 2)$

4, because $(40 \div 4 = 10)$ 5, because the last digit is 0

8, because of 2 and 4   10, because the last digit is 0

Then 240 is divisible by 2, 3, 4, 5, 8, 10

PRACTICES:

Use the divisibility rules to find the factors of each number.

1) 12	2 3 4 5 6 7 8 9 10	2) 24	2 3 4 5 6 7 8 9 10
3) 36	2 3 4 5 6 7 8 9 10	4) 18	2 3 4 5 6 7 8 9 10
5) 30	2 3 4 5 6 7 8 9 10	6) 54	2 3 4 5 6 7 8 9 10
7) 90	2 3 4 5 6 7 8 9 10	8) 80	2 3 4 5 6 7 8 9 10
9) 72	2 3 4 5 6 7 8 9 10	10) 84	2 3 4 5 6 7 8 9 10

Score: ....

Answer Key

1) 12	**2** **3** **4** 5 **6** 7 8 9 10	2) 24 **2** **3** **4** 5 **6** 7 **8** 9 10
3) 36	**2** **3** **4** 5 **6** 7 8 **9** 10	4) 18 **2** **3** 4 5 **6** 7 8 **9** 10
5) 30	**2** 3 4 **5** **6** 7 8 9 **10**	6) 54 **2** **3** 4 5 **6** 7 8 **9** 10
7) 90	**2** **3** 4 **5** **6** 7 8 **9** **10**	8) 80 **2** 3 **4** **5** 6 7 **8** 9 **10**
9) 72	**2** **3** **4** 5 **6** 7 **8** **9** 10	10) 84 **2** **3** **4** 5 **6** **7** 8 9 10

Name: ....

Adding and Subtracting Fractions

✓ Find equivalent fractions with the equivalent divisor before you can add or subtract fractions with totally different divisors.

✓ Adding and Subtracting with the equivalent divisors:

$$\frac{a}{b} + \frac{c}{b} = \frac{a+c}{b} \ , \frac{a}{b} - \frac{c}{b} = \frac{a-c}{b}$$

✓ Adding and Subtracting fractions with different divisors:

$$\frac{a}{b} + \frac{c}{d} = \frac{ad+cb}{bd} \ , \frac{a}{b} - \frac{c}{d} = \frac{ad-cb}{bd}$$

EXAMPLE:

Subtract fractions. $\frac{2}{3} - \frac{1}{2} = ?$

For "unlike" fractions, find equivalent fractions with the same divisors before you can add or subtract fractions with different divisors. Use this formula: $\frac{a}{b} - \frac{c}{d} = \frac{ad-cb}{bd}$

$$\frac{2}{3} - \frac{1}{2} = \frac{(2)(2) - (1)(3)}{3 \times 2} = \frac{4-3}{6} = \frac{1}{6}$$

PRACTICES:

Add fractions.	Subtract fractions.
1) $\frac{1}{4} + \frac{2}{3}$	2) $\frac{1}{2} - \frac{1}{5}$
3) $\frac{1}{3} + \frac{1}{2}$	4) $\frac{1}{7} - \frac{1}{9}$
5) $\frac{1}{4} + \frac{5}{7}$	6) $\frac{3}{5} - \frac{1}{15}$
7) $\frac{6}{7} + \frac{3}{21}$	8) $\frac{1}{3} - \frac{1}{4}$
9) $\frac{5}{13} + \frac{1}{2}$	10) $\frac{6}{5} - \frac{5}{6}$

Score: ....

Answer Key	
1) $\frac{11}{12}$	2) $\frac{3}{10}$
3) $\frac{5}{6}$	4) $\frac{2}{63}$
5) $\frac{27}{28}$	6) $\frac{8}{15}$
7) 1	8) $\frac{1}{12}$
9) $\frac{23}{26}$	10) $\frac{11}{30}$

Name: ....

Multiplying and Dividing Fractions

- ✓ How to multiply fractions: multiply the top numbers and multiply the bottom numbers.
- ✓ How to change fractions: Keep, Change, Flip
- ✓ Keep the first fraction then change division sign into multiplication sign and flip the numerator and denominator of the second fraction. Then, solve it

EXAMPLE:

Multiplying fractions. $\frac{5}{6} \times \frac{3}{4} =$

Multiply the upper numbers and multiply the lower numbers.

$\frac{5}{6} \times \frac{3}{4} = \frac{5 \times 3}{6 \times 4} = \frac{15}{24}$, simplify: $\frac{15}{24} = \frac{15 \div 3}{24 \div 3} = \frac{5}{8}$

Dividing fractions. $\frac{1}{4} \div \frac{2}{3} =$

Keep the first fraction then change division sign into multiplication sign and flip the numerator and denominator of the second fraction. Then: $\frac{1}{4} \times \frac{3}{2} = \frac{1 \times 3}{4 \times 2} = \frac{3}{8}$

PRACTICES:

Multiplying fractions. Then simplify.	Dividing fractions.
1) $\frac{3}{5} \times \frac{5}{9}$	2) $\frac{4}{9} \div 4$
3) $\frac{5}{21} \times \frac{7}{10}$	4) $\frac{32}{25} \div \frac{8}{5}$
5) $\frac{3}{29} \times \frac{29}{3}$	6) $\frac{2}{7} \div \frac{8}{35}$
7) $\frac{8}{11} \times 11$	8) $\frac{12}{25} \div \frac{3}{5}$
9) $\frac{7}{9} \times \frac{12}{28}$	10) $7 \div \frac{2}{3}$

Score: ....

Answer Key	
1) $\frac{1}{3}$	2) $\frac{1}{9}$
3) $\frac{1}{6}$	4) $\frac{4}{5}$
5) 1	6) $1\frac{1}{4}$
7) 8	8) $\frac{4}{5}$
9) $\frac{1}{3}$	10) $10\frac{1}{2}$

Name: ....

Adding Mixed Numbers

Use these steps for both adding and subtracting mixed numbers.
- ✓ Add whole number of the mixed numbers.
- ✓ Add the fractions of every mixed number.
- ✓ Find the Least Common Divisor (LCD) if needed.
- ✓ Add whole numbers and fractions.
- ✓ Write your answer in simplest form.

EXAMPLE:

$1\frac{3}{4} + 2\frac{3}{8} = ?$

Rewriting our equation with parts separated, $1 + \frac{3}{4} + 2 + \frac{3}{8}$, Solving the whole number

parts $1 + 2 = 3$, Solving the fraction parts $\frac{3}{4} + \frac{3}{8}$, and rewrite to solve with the equivalent

fractions.

$\frac{6}{8} + \frac{3}{8} = \frac{9}{8} = 1\frac{1}{8}$, then combining the whole and fraction parts $3 + 1 + \frac{1}{8} = 4\frac{1}{8}$

PRACTICES:

Add.

1) $1\frac{1}{7} + 2\frac{1}{3}$	2) $1\frac{1}{2} + 3\frac{2}{3}$
3) $1\frac{2}{5} + 2\frac{1}{10}$	4) $7 + 2\frac{1}{2}$
5) $4\frac{1}{3} + 2\frac{2}{3}$	6) $2\frac{2}{3} + 1\frac{1}{4}$
7) $2\frac{3}{4} + 3\frac{1}{8}$	8) $9 + 1\frac{1}{9}$
9) $4\frac{5}{12} + 2\frac{3}{4}$	10) $3\frac{1}{7} + 2\frac{3}{14}$

Score: ....

Answer Key	
1) $3\frac{10}{21}$	2) $5\frac{1}{6}$
3) $3\frac{1}{2}$	4) $9\frac{1}{2}$
5) 7	6) $3\frac{11}{12}$
7) $5\frac{7}{8}$	8) $10\frac{1}{9}$
9) $7\frac{1}{6}$	10) $5\frac{5}{14}$

Name: ....

Subtracting Mixed Numbers

Use these steps for both adding and subtracting mixed numbers.
- ✓ From whole number of the first mixed number, subtract the whole number of second mixed number.
- ✓ From first fraction subtract the second.
- ✓ Find the Least Common Divisor (LCD) if needed.
- ✓ Add the result of whole numbers and fractions.
- ✓ Write your answer in simplest terms.

EXAMPLE:

$5\frac{2}{3} - 2\frac{1}{4} = ?$

Rewriting our equation with parts separated, $5 + \frac{2}{3} - 2 - \frac{1}{4}$

Solving the whole number parts $5 - 2 = 3$, Solving the fraction parts, $\frac{2}{3} - \frac{1}{4} = \frac{8-3}{12} = \frac{5}{12}$

Joining the whole and fraction parts, $3 + \frac{5}{12} = 3\frac{5}{12}$

PRACTICES:

Subtract.

1) $5\frac{2}{7} - 2\frac{1}{14}$	2) $4\frac{2}{5} - \frac{2}{3}$
3) $3\frac{3}{7} - 1\frac{1}{14}$	4) $7\frac{2}{5} - 5\frac{1}{3}$
5) $4\frac{1}{2} - 2\frac{4}{8}$	6) $11\frac{5}{12} - 8\frac{3}{4}$
7) $7\frac{5}{12} - 5\frac{7}{12}$	8) $5\frac{2}{9} - 2\frac{1}{18}$
9) $3\frac{2}{5} - 2\frac{1}{5}$	10) $3\frac{4}{9} - 1\frac{2}{9}$

Score: ....

Answer Key	
1) $3\frac{3}{14}$	2) $3\frac{11}{15}$
3) $2\frac{5}{14}$	4) $2\frac{1}{15}$
5) 2	6) $2\frac{2}{3}$
7) $1\frac{5}{6}$	8) $3\frac{1}{6}$
9) $1\frac{1}{5}$	10) $2\frac{2}{9}$

Name: ....

Multiplying Mixed Numbers

✓ Convert the mixed numbers into improper fractions. (Improper fraction is a fraction in which the numerator is greater than denominator)

✓ Multiply fractions and write in simplest form if needed.

$$a\frac{c}{b} = a + \frac{c}{b} = \frac{ab+c}{b}$$

EXAMPLE:

Multiply mixed numbers. $4\frac{3}{5} \times 2\frac{1}{3} = ?$

Changing mixed numbers to fractions, $\frac{23}{5} \times \frac{7}{3}$, Applying the fractions formula for

multiplication, $\frac{23 \times 7}{5 \times 3} = \frac{161}{15} = 10\frac{11}{15}$

PRACTICES:

Find each product.

1) $2\frac{1}{3} \times \frac{1}{2}$	2) $1\frac{2}{5} \times \frac{2}{3}$
3) $2\frac{4}{3} \times 2\frac{2}{6}$	4) $2\frac{1}{2} \times 1\frac{2}{4}$
5) $3\frac{1}{2} \times 1\frac{2}{3}$	6) $1\frac{1}{7} \times 1\frac{3}{4}$
7) $1\frac{1}{4} \times 2\frac{6}{5}$	8) $3\frac{1}{2} \times 4\frac{2}{5}$
9) $1\frac{2}{5} \times 2\frac{1}{3}$	10) $5\frac{7}{12} \times 2\frac{4}{9}$

Score: ....

Answer Key	
1) $1\frac{1}{6}$	2) $\frac{14}{15}$
3) $7\frac{7}{9}$	4) $3\frac{3}{4}$
5) $5\frac{5}{6}$	6) 2
7) 4	8) $15\frac{2}{5}$
9) $3\frac{4}{15}$	10) $13\frac{35}{54}$

Name: ....

Dividing Mixed Numbers

✓ Change the mixed numbers into improper fractions.

✓ Divide fractions and write in simplest form if needed.

$$a\frac{c}{b} = a + \frac{c}{b} = \frac{ab + c}{b}$$

EXAMPLE:

Find the quotient. $2\frac{1}{2} \div 1\frac{1}{5} = ?$

Changing mixed numbers to fractions, $\frac{5}{2} \div \frac{6}{5}$, Applying the fractions formula for

multiplication, $\frac{5}{2} \times \frac{5}{6} = \frac{5 \times 5}{2 \times 6} = \frac{25}{12} = 2\frac{1}{12}$

PRACTICES:

Find each quotient.

1) $2\frac{3}{5} \div 1\frac{3}{8}$

2) $\frac{3}{2} \div 2\frac{3}{4}$

3) $1\frac{4}{7} \div 2\frac{2}{3}$

4) $1\frac{2}{3} \div 2\frac{1}{3}$

5) $0 \div 4\frac{2}{5}$

6) $2\frac{2}{5} \div 1\frac{1}{2}$

7) $1\frac{2}{3} \div 2\frac{1}{5}$

8) $3\frac{2}{7} \div 4\frac{3}{5}$

9) $1\frac{1}{4} \div 2\frac{4}{5}$

10) $2 \div 3\frac{1}{3}$

Score: ....

Answer Key

1) $1\frac{49}{55}$	2) $\frac{6}{11}$
3) $\frac{33}{56}$	4) $\frac{5}{7}$
5) 0	6) $1\frac{3}{5}$
7) $\frac{25}{33}$	8) $\frac{5}{7}$
9) $\frac{25}{56}$	10) $\frac{3}{5}$

Name: ....

Comparing Decimals

Decimal is a fraction written in a unique form. For example, instead of writing $\frac{1}{2}$ you can write as **0.5**.

For comparison of decimals:

✓ Compare every digit of two decimals in the same place value.

✓ Start from left. Compare ones, tens, hundreds, tenths, hundredths, etc.

✓ To compare numbers, use these symbols:

- Equal to =,   Less than <,   Greater than >

- Greater than or equal ≥,   Less than or equal ≤

EXAMPLE:

Compare **0.20** and **0.02**.

0.20 *is greater than* 0.02, because the tenth place of 0.20 is 2, but the tenth place of 0.02 is zero. Then: 0.20 > 0.02

PRACTICES:

Write the correct comparison symbol (>, < or =).

1) 0.025 _____ 0.25	2) 0.9 _____ 0.888
3) 4.510 _____ 4.150	4) 10.01 _____ 10.10
5) 0.987 _____ 0.991	6) 18.004 _____ 18.040
7) 0.020 _____ 0.20	8) 0.071___ _0.700
9) 0.08____0.009	10) 0.690 _____ 0.609

Score: ....

Answer Key	
1) <	2) >
3) >	4) <
5) <	6) <
7) <	8) <
9) >	10) >

Name: ....

Rounding Decimals

✓ To round a decimal, you must find the place value you'll round to.

✓ Then, find the digit to the right of the place value you're rounding to.

- If it is 5 or greater, add 1 to the place value you're rounding to and remove all digits on its right side.

- If the digit to the right of the place value is smaller than 5, keep the place value and remove all digits on the right.

EXAMPLE:

Round 2. 1837 to the thousandth-place value.

First have a look at the next place value to the right, (tens thousandths). It's 7 and it's found to be greater than 5. So, add 1 to the digit in the thousandth place.

Thousandth place is 3. $\rightarrow 3 + 1 = 4$, then, the answer is 2.184

PRACTICES:

Round each decimal number to the nearest place indicated.

1) 6.08	2) 12.267
3) 9.301	4) 10.071
5) 55.89	6) 59.15
7) 329.018	8) 92.410
9) 1.499	10) 25.621

Score: ....

Answer Key	
1) 6.1	2) 12.3
3) 9.3	4) 10.07
5) 56	6) 59
7) 330	8) 92.4
9) 1.5	10) 26

Name: ....

Adding and Subtracting Decimals

- ✓ Arrange the numbers in line.
- ✓ Add zeros to have same number of digits for both the numbers.
- ✓ Add or subtract by using column subtraction or addition.

EXAMPLE:

Add. $2.5 + 1.24 =$

First line up the numbers: $\begin{array}{r} 2.5 \\ +1.24 \\ \hline \end{array}$ → Add zeros to have same number of digits for both

numbers. $\begin{array}{r} 2.50 \\ +1.24 \\ \hline \end{array}$, Start with the hundredths place. $0 + 4 = 4$, $\begin{array}{r} 2.50 \\ +1.24 \\ \hline 4 \end{array}$, Continue with tenths

place. $5 + 2 = 7$, $\begin{array}{r} 2.50 \\ +1.24 \\ \hline .74 \end{array}$. Add the ones place. $2 + 1 = 3$, $\begin{array}{r} 2.50 \\ +1.24 \\ \hline 3.74 \end{array}$

Subtract decimals. $4.67 - 2.15 =$ $\begin{array}{r} 4.67 \\ -2.15 \\ \hline \end{array}$

Start with the hundredths place. $7 - 5 = 2$, $\begin{array}{r} 4.67 \\ -2.15 \\ \hline 2 \end{array}$, continue with tenths place. $6 - 1 = 5$

$\begin{array}{r} 4.67 \\ -2.15 \\ \hline .52 \end{array}$, subtract the ones place. $4 - 2 = 2$, $\begin{array}{r} 4.67 \\ -2.15 \\ \hline 2.52 \end{array}$.

PRACTICES:

Add and subtract decimals.

1) $\begin{array}{r} 87.15 \\ -\ 32.35 \\ \hline \end{array}$	2) $\begin{array}{r} 90.43 \\ +\ 44.09 \\ \hline \end{array}$
3) $\begin{array}{r} 58.56 \\ +\ 12.10 \\ \hline \end{array}$	4) $\begin{array}{r} 65.23 \\ -\ 56.48 \\ \hline \end{array}$
5) $\begin{array}{r} 98.125 \\ +\ 58.54 \\ \hline \end{array}$	6) $\begin{array}{r} 162.05 \\ -\ 83.65 \\ \hline \end{array}$

Solve.

7) $\_\_\_ + 5.0 = 9.08$	8) $7.06 + \_\_\_ = 24.6$
9) $21.9 - \_\_\_ = 6.9$	10) $32.12 - \_\_\_ = 12.07$

Score: ....

Answer Key	
1) 54.8	2) 134.52
3) 70.66	4) 8.75
5) 156.665	6) 78.4
7) 4.08	8) 17.54
9) 15	10) 20.05

Name: ....

Multiplying Decimals

- ✓ Arrange and multiply the numbers as you do with whole numbers.
- ✓ Then count the total number of decimal places in every factor.
- ✓ Place the decimal point in the product.

EXAMPLE:

Find the product. $0.50 \times 0.20 =$

Arrange and multiply the numbers as you do with whole numbers. Line up the numbers:

$\begin{array}{r} 50 \\ \times 20 \\ \hline \end{array}$, Start with the ones place $\to 0 \times 50 = 0$, $\begin{array}{r} 50 \\ \times 20 \\ \hline 0 \end{array}$, Continue with other digits $\to$

$2 \times 50 = 100$, $\begin{array}{r} 50 \\ \times 20 \\ \hline 1,000 \end{array}$, Count the total number of decimal places in both of the factors

(4). Then Place the decimal point in the product.

Then: $\begin{array}{r} 0.50 \\ \times 0.20 \\ \hline 0.1000 \end{array}$ $\to 0.50 \times 0.20 = 0.1$

PRACTICES:

Find each product.

1) $\begin{array}{r} 1.5 \\ \times 0.16 \\ \hline \end{array}$	2) $\begin{array}{r} 5.3 \\ \times 1.9 \\ \hline \end{array}$
3) $\begin{array}{r} 0.06 \\ \times 2.5 \\ \hline \end{array}$	4) $\begin{array}{r} 3.19 \\ \times 21.5 \\ \hline \end{array}$
5) $\begin{array}{r} 9.3 \\ \times 11.5 \\ \hline \end{array}$	6) $\begin{array}{r} 3.01 \\ \times 2.1 \\ \hline \end{array}$
7) $\begin{array}{r} 5.0 \\ \times 1.4 \\ \hline \end{array}$	8) $\begin{array}{r} 23.8 \\ \times 10 \\ \hline \end{array}$
9) $\begin{array}{r} 21.5 \\ \times 0.001 \\ \hline \end{array}$	10) $\begin{array}{r} 8.21 \\ \times 3.1 \\ \hline \end{array}$

Score: ....

Answer Key	
1) 0.24	2) 10.07
3) 0.15	4) 68.585
5) 106.95	6) 6.321
7) 7	8) 238
9) 0.0215	10) 25.451

Name: ....

Dividing Decimals

- ✓ If the divisor is not a whole number, transfer decimal point to right to make it a whole number. Do the same step for dividend.
- ✓ Divide same to whole numbers.

EXAMPLE:

Find the quotient. $1.20 \div 0.2 =$

The divisor is not a whole number. Multiply it by 10 to get 2. Do the same step for the dividend to get 12. Now, divide: $12 \div 2 = 6$. The answer is 6.

PRACTICES:

Find each quotient.

1) $25.7 \div 0.5$	2) $67.2 \div 4$
3) $61.75 \div 1.9$	4) $18.0 \div 1.2$
5) $12.4 \div 10$	6) $2.2 \div 100$
7) $1.88 \div 100$	8) $55.1 \div 100$
9) $0.1 \div 100$	10) $0.25 \div 10$

Score: ....

Answer Key	
1) 51.4	2) 16.8
3) 32.5	4) 15
5) 1.24	6) 0.022
7) 0.0188	8) 0.551
9) 0.001	10) 0.025

Name: ....

Converting Between Fractions, Decimals and Mixed

How to convert fraction into Decimal:

✓ Divide the numerator by denominator.

How to convert decimal into Fraction:

✓ Write decimal over 1.

✓ Multiply both numerator value and denominator value by 10 for each digit on the right side of the decimal point.

✓ Make it to simplest form.

EXAMPLE:

What is long division of $\frac{5}{8}$ =?

In that case we put extra zeros and did $\frac{5.000}{8}$ to get 0.625

PRACTICES:

Convert fractions to decimals.	Convert decimal into fraction or mixed numbers
1) $\frac{4}{10}$	2) 3.6
3) $\frac{3}{8}$	4) 0.07
5) $\frac{4}{12}$	6) 0.15
7) $\frac{5}{16}$	8) 2.7
9) $\frac{60}{100}$	10) 2.5

Score: ....

Answer Key	
1) 0.4	2) $3\frac{3}{5}$
3) 0.375	4) $\frac{7}{100}$
5) 0.333	6) $\frac{3}{20}$
7) 0.3125	8) $2\frac{7}{10}$
9) 0.6	10) $2\frac{1}{2}$

Chapter 3 : Proportion, Ratio, Percent

Topics that you'll learn in this chapter:

➢ Writing and Simplifying Ratios

➢ Create a Proportion

➢ Similar Figures

➢ Simple Interest

➢ Ratio and Rates Word Problems

➢ Percentage Calculations

➢ Converting Between Percent, Fractions, and Decimals

➢ Percent Problems

➢ Markup, Discount, and Tax

"Do not worry about your difficulties in mathematics. I can assure you mine are still greater." – **Albert Einstein**

Name: ....

Writing Ratios

✓ A ratio is a comparison of two numbers, and it can be written as a division.

EXAMPLE:

$3:5 =?$

Both numbers 3 and 5 are divisible by 8 , $\Rightarrow 3 \div 8 = \frac{3}{8}, 5 \div 8 = \frac{5}{8},$

Then: $3:5 = \frac{3}{8}$ and $\frac{5}{8}$.

PRACTICES:

Express each ratio as a rate and unite rate.	Express each ratio as a fraction in the simplest form
1) 80 dollars for 4 chairs.	2) 13 cups to 39 cups.
3) 125 miles on 25 gallons of gas.	4) 17 cakes out of 51 cakes
5) 147 miles on 7 hours.	6) 35 red desks out of 125 desks
7) 12 inches of snow in 24 hours.	8) 8 story books out of 32 books
9) 14 dimes to 112 dimes.	10) 12 gallons to 20 gallons

Score: ....

Answer Key

1) $\frac{80 \text{ dollars}}{4 \text{ books}}$, 20.00 dollars per chair

2) $\frac{1}{3}$

3) $\frac{125 \text{ miles}}{25 \text{ gallons}}$, 5 miles per gallon

4) $\frac{1}{3}$

5) $\frac{147 \text{ miles}}{7 \text{ hours}}$, 21 miles per hour

6) $\frac{7}{25}$

7) $\frac{12" \text{ of snow}}{24 \text{ hours}}$, 0.5 inches of snow per hour

8) $\frac{1}{4}$

9) $\frac{14 \text{ dimes}}{112 \text{ dimes}}$, $\frac{1}{8}$ per dime

10) $\frac{3}{5}$

Name: ....

Simplifying Ratios

- ✓ Ratios are used to compare two numbers.

- ✓ Ratios can be written as a fraction, using colon or the word "to".

- ✓ You can calculate identical ratios by multiplying or dividing both sides of the ratio by the same number.

EXAMPLE:

Simplify. $8:4 =$

Both numbers 8 and 4 are divisible by 4 , $\Rightarrow 8 \div 4 = 2, 4 \div 4 = 1,$

Then: $8:4 = 2:1$

PRACTICES:

Reduce each ratio.

1) 49: 14	2) 22: 55
3) 35: 25	4) 18: 99
5) 16: 36	6) 64: 72
7) 4: 60	8) 70: 40
9) 8: 64	10) 16: 24

Score: ....

Answer Key	
1) 7: 2	2) 2: 5
3) 7: 5	4) 2: 11
5) 4: 9	6) 8: 9
7) 1: 15	8) 7: 4
9) 1: 8	10) 2: 3

Name: ....

Create a Proportion

- ✓ A proportion carries two equal fractions! A proportion means equality of two fractions.
- ✓ If you want to create a proportion, simply find (or create) two equal fractions.

EXAMPLE:

Explain if these ratios form a proportion. $\frac{3}{5}$ and $\frac{24}{45}$

Use cross multiplication: $\frac{3}{5} = \frac{24}{45} \rightarrow 3 \times 45 = 5 \times 24 \rightarrow 135 = 120$, which is not correct.

Thus, this pair of ratios doesn't form a proportion.

PRACTICES:

Create proportion from the given set of numbers.

1) 3, 2, 9, 6	2) 4, 18, 12, 6
3) 5, 11, 25, 55	4) 24, 7, 21, 8
5) 49, 7, 12, 84	6) 15, 12, 30, 24
7) 20, 10, 200, 1	8) 9, 27, 81, 3
9) 4, 2, 16, 32	10) 9, 6, 27, 18

Score: ....

Answer Key	
1) 2: 6 = 3: 9	2) 4: 12 =6: 18
3) 5: 25 = 11: 55	4) 8: 24 =7: 21
5) 7: 49 = 12: 84	6) 12: 24 =15: 30
7) 1: 10 = 20: 200	8) 3: 27 =9: 81
9) 2: 16=4: 32	10) 6: 18 = 9: 27

Name: ....

Similar Figures

✓ Two or more figures are equivalent if their corresponding angles are equal, and the corresponding sides are in proportion.

EXAMPLE:

4–5–6 triangle is like an 8–10–12 triangle.

PRACTICES:

Each pair of figures is similar. Find the missing side.

1)
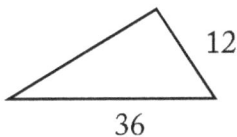 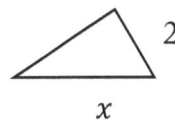

2)
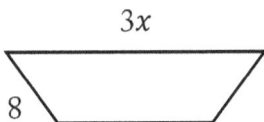 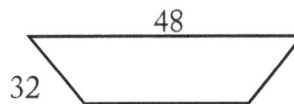

3)
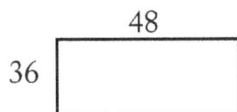 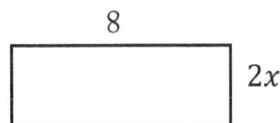

4)
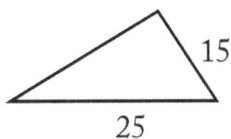 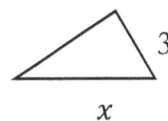

5)
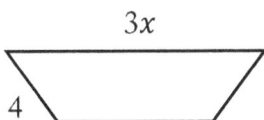 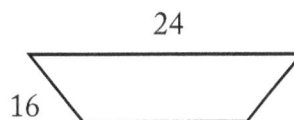

6)	2x / 4	36 / 24
7)	63 / 54	7 / 6x
8)	27 / 72	3 / x
9)	4 / 6	2x / 60
10)	45 / x	5 / 2

Score: ....

Answer Key	
1) 6	2) 4
3) 3	4) 5
5) 2	6) 3
7) 1	8) 8
9) 20	10) 18

Name: ....

Ratio and Rates Word Problems

✓ To solve a rate word problem or a ratio, create a proportion and then use cross multiplication method.

EXAMPLE:

A tree **32 feet** tall has a shadow **12 feet** long. Jack is **6 feet** tall. How long is Jack's shadow?

To solve for the missing number, write in a proportion.

$\frac{32}{12} = \frac{6}{x} \rightarrow 32x = 6 \times 12 = 72$

$32x = 72 \rightarrow x = \dfrac{72}{32} = 2.25$

PRACTICES:

Solve.

1) In a party, 8 soft drinks are required for every 35 guests. If there are 560 guests, how many soft drinks is required?

2) You can buy 6 cans of green beans at a supermarket for $3.50. How much does it cost to buy 42 cans of green beans?

3) The price of 5 bananas at the first Market is $1.05. The price of 7 of the same bananas at second Market is $1.07. Which place is the better buy?

4) In Peter's class, 21 of the students are tall and 9 are short. In Elise's class 56 students are tall and 24 students are short. Which class has a higher ratio of tall to short students?

5) The bakers at a Bakery can make 110 bagels in 4 hours. How many bagels can they bake in 6 hours? What is that rate per hour?

6) A certain sweet recipe calls for 3 kg of sugar for every 6 kg of flour. If 63 kg of this sweet must be prepared, how much sugar is required?

7) In a mixture of 45 liters, the ratio of sugar solution to salt solution is 1:2. What is the amount of sugar solution to be added if the ratio must be 2:1?

8) In a bag of red and green sweets, the ratio of red sweets to green sweets is 3:4. If the bag contains 120 green sweets, how many red sweets are there?

9) If the ratio of chocolates to ice-cream cones in a box is 5:8 and the number of chocolates is 30, find the number of ice-cream cones.

10) In a group, the ratio of doctors to lawyers is 5:4. If the total number of people in the group is 72, what is the number of lawyers in the group?

Score: ....

Answer Key	
1) 128	2) $24.5
3) The price at the second Market is a better buy.	4) The ratio for both classes equal 7 to 3.
5) 165, the rate is 27.5 per hour.	6) 21 kg (3+6=9, $\frac{63}{9} = 7$. Therefore, 3:6=21:42)
7) 45	8) 90
9) 48	10) 32

Name: ....

Percentage Calculations

- ✓ Percent is called the ratio of a number and 100. It always possesses the same denominator, 100. The symbol used for percent is %.
- ✓ Percent is another method to write decimals or fractions. For example:

$$40\% = 0.40 = \frac{40}{100} = \frac{2}{5}$$

- ✓ Use the given formula to find part, whole, or percent:

$$\text{part} = \frac{\text{percent}}{100} \times \text{whole}$$

EXAMPLE:

What is **10%** of **45**?

Use this formula: $\text{part} = \frac{\text{percent}}{100} \times \text{whole}$

$$\text{part} = \frac{10}{100} \times 45 \rightarrow \text{part} = \frac{1}{10} \times 45 \rightarrow \text{part} = \frac{45}{10} \rightarrow \text{part} = 4.5$$

PRACTICES:

Calculate the percentages.

1) 75% of 45	2) 50% of 66
3) 90% of 58	4) 25% of 88
5) 5% of 100	6) 80% of 60

Solve.

7) What percentage of 60 is 6	8) 6.76 is what percentage of 52?
9) 17 is what percentage of 85?	10) Find what percentage of 96 is 24.

Score: ....

Answer Key	
1) 33.75	2) 33
3) 52.2	4) 22
5) 5	6) 48
7) 10%	8) 13%
9) 20%	10) 25%

Name: ....

Percent Problems

✓ In each percent question, we are finding the base, or part or the percent.
✓ Use the following equations to find each missing portion.
- Base = Part ÷ Percent
- Part = Percent × Base
- Percent = Part ÷ Base

EXAMPLE:

20 is 5% of what number?

Use the formula: $Base = Part \div Percent \rightarrow Base = 20 \div 0.05 = 400$

20 is 5% of 400

PRACTICES:

Solve each problem.

1) 52% of what number is 13?	2) What is 15% of 9 inches?
3) What percent of 185.6 is 23.2?	4) 24 is 72% of what?
5) 35 is what percent of 70?	6) 10 is 200% of what?
7) 14 is what percent of 70?	8) 26% of 100 is what number?

9) Mia requires 50% to pass. If she gets 250 marks and falls short by 90 marks, what were the maximum marks she could have got?

10) Jack scored 14 out of 70 marks in mathematics, 9 out of 10 marks in history and 56 out of 100 marks in science. In which subject his percentage of marks is the best?

Score: ....

Answer Key	
1) 25	2) 1.35
3) 12.5	4) 33.33
5) 50%	6) 5
7) 20%	8) 26
9) 680	10) History

Name: ....

Markup, Discount, and Tax

- ✓ Markup = selling price – cost
- ✓ Markup rate = markup is divided by the cost
- ✓ Discount = Multiply the rate of discount by regular price.
- ✓ Tax: To find tax, multiply the taxable amount (income, property value, etc.) to the tax rate.
- ✓ To find tip, multiply selling price to the rate.

EXAMPLE:

With an **10%** discount, Ella was able to save **$20** on a dress. What was the original price of the dress?

$$10\% \; of \; x = \; 20, \frac{10}{100} \times x = \; 20, x = \frac{100 \times 20}{10} = 200$$

PRACTICES:

Find the selling price of each item.

1) Cost of a chair: $20, markup: 30%, discount: 10%, tax: 10%

2) Cost of computer: $1,600.00, markup: 65%

3) Cost of a pen: $3.20, markup: 50%, discount: 15%, tax: 5%

4) Cost of a puppy: $1,800, markup: 40%, discount: 10%

5) Cost of a book: $50, markup: 40%, discount: 20%, tax: 5%

6) Original price of a tablet: $400, discount: 20% Tax: 5%,

7) Original price of a book: $50, markup:20% Discount: 20%, Tax: 2.5%,

8) Original price of a cellphone: $500, markup:14% Discount: 25%, Tax: 1.6%,

9) Original price of a sofa: $800, markup:10% Discount: 15%, Tax: 1.5%,

10) Original price of a car: $40,000, markup:12% Discount: 25%, Tax: 6.5%,

Score: ....

Answer Key	
1) $25.74	2) $2,640
3) $4.284	4) $2,268
5) $58.8	6) $336
7) $49.2	8) $434.34
9) $759.22	10) $35,784

Name: ....

Simple Interest

- ✓ Simple Interest: The charge for borrowing money or the return for lending it.

  To solve a simple interest problem, use this formula:

- ✓ Interest = principal × rate × time $\Rightarrow I = p \times r \times t$

EXAMPLE:

Find simple interest for $5,200 at 4% for 3 years.

Use Interest formula: $I = prt$

$P = \$5,200, r = 4\% = \frac{4}{100} = 0.04$ and $t = 3$

Then: $I = 5,200 \times 0.04 \times 3 = \$624$

PRACTICES:

Use simple interest to find the ending balance.

1) $1,200 at 15% for 3 years.	2) $320,000 at 2.85% for 7 years.
3) $1,500 at 2.25% for 12 years.	4) $12,500 at 6.2% for 4 years.
5) $31,000 at 1.5% for 10 months.	6) $18,000 at 5.2% for 5 years.

7) Emily puts $6,000 into an investment yielding 3.25% annual simple interest; she left the money in for 3 years. How much interest does Sara get at the end of those 3 years?

8) A new car, valued at $42,000, depreciates at 7.5% per year from original price. Find the value of the car 6 years after purchase.

9) $880 interest is earned on a principal of $2,200 at a simple interest rate of 4% interest per year. For how many years was the principal invested?

10) A bank is offering 3.2% simple interest on a savings account. If you deposit $15,000, how much interest will you earn in six years?

Score: ....

Answer Key	
1) $1,740	2) $383,840.00
3) $1,905.00	4) $15,600
5) $31,387.50	6) $22,680
7) $585.00	8) $23,100
9) 10 years	10) $2,880

Name: ....

Converting Between Percent, Fractions, and Decimals

✓ To a percent: We move the decimal point 2 places to the right and add the percentage (%) symbol.

✓ Divide by 100 to change a number from percent to decimal.

EXAMPLE:

$30\% = 0.30$

$0.24 = 24\%$

PRACTICES:

Converting fractions to decimals.	Write each decimal as a percent.
1) $\frac{23}{10}$	2) 0.002
3) $\frac{2}{20}$	4) 0.08
5) $\frac{7}{100}$	6) 0.2
7) $\frac{20}{50}$	8) 3.25
9) $\frac{3}{60}$	10) 1.01

Score: ....

Answer Key	
1) 2.3	2) 0.2%
3) 0.1	4) 8%
5) 0.07	6) 20%
7) 0.4	8) 325%
9) 0.05	10) 101%

Chapter 4 : Exponents and Radicals

Topics that you'll learn in this chapter:

- ➢ Multiplication Property of Exponents

- ➢ Division Property of Exponents

- ➢ Powers of Products and Quotients

- ➢ Zero, Negative Exponents and Bases

"Mathematics is no more computation than typing is literature." – *John Allen Paulos*

Name: ....

Multiplication Property of Exponents

- ✓ Exponents are shorthand for recurrent multiplication of the identical number by itself. For example, instead of writing 2×2, we can write 2^2. For $3 \times 3 \times 3 \times 3$, we can write 3^4

- ✓ In algebra, a variable is a letter used as a replacement for a number. The most common letters are: $x, y, z, a, b, c, m,$ and n.

- ✓ Exponent's rules: $(x^a)^b = x^{a \times b}$, $(xy)^a = x^a \times y^a$,

  $x^a \times x^b = x^{a+b}$, $x^a \times y^a = (xy)^a$,

EXAMPLE:

Multiply. $-2x^5 \times 7x^3 =$

Use Exponent's rules: $x^a \times x^b = x^{a+b} \rightarrow x^5 \times x^3 = x^{5+3} = x^8$

Then: $-2x^5 \times 7x^3 = -14x^8$

PRACTICES:

Simplify.

1) $4^3 \times 4^2$	2) $2 \times 2^2 \times 2^3$
3) $2^4 \times 2$	4) $8x^2 \times x$
5) $15x^7 \times x$	6) $3x \times x^3$
7) $2x^5 \times 5x^4$	8) $5x^2 \times 3x^2y^2$
9) $6y^5 \times 8xy^2$	10) $5xy^3 \times 4x^3y^2$

Score: ....

Answer Key	
1) 4^5	2) 2^6
3) 2^5	4) $8x^3$
5) $15x^8$	6) $3x^4$
7) $10x^9$	8) $15x^4y^2$
9) $48xy^7$	10) $20x^4y^5$

Name: ....

Division Property of Exponents

✓ For division of exponents, we can use these formulas: $\left(\frac{a}{b}\right)^c = \frac{a^c}{b^c}$, $b \neq 0$

$$\frac{x^a}{x^b} = x^{a-b}, x \neq 0 \qquad \frac{x^a}{y^a} = \left(\frac{x}{y}\right)^a, y \neq 0$$

$$\frac{x^a}{x^b} = \frac{1}{x^{b-a}}, x \neq 0, \qquad \frac{1}{x^b} = x^{-b}$$

EXAMPLE:

Simplify. $\frac{4x^3y}{36x^2y^3} =$

First you cancel the common factor: $4 \rightarrow \frac{4x^3y}{36x^2y^3} = \frac{x^3y}{9x^2y^3}$

Use Exponent's rules: $\frac{x^a}{x^b} = x^{a-b} \rightarrow \frac{x^3}{x^2} = x^{3-2}$

Then: $\frac{4x^3y}{36x^2y^3} = \frac{xy}{9y^3} \rightarrow$ now cancel the common factor: $y \rightarrow \frac{xy}{9y^3} = \frac{x}{9y^2}$

PRACTICES:

Simplify.

1) $\frac{4^3}{4}$	2) $\frac{51}{51^{14}}$
3) $\frac{5^2}{5^3}$	4) $\frac{3^4}{15^4}$
5) $\frac{x}{x^7}$	6) $\frac{42x^2}{6x^2}$
7) $\frac{3x^{-3}}{12x^{-1}}$	8) $\frac{81x^5}{9x^3}$
9) $\frac{3x^4}{4x^5}$	10) $\frac{21x}{3x^2}$

Score: ....

Answer Key	
1) 4^2	2) $\frac{1}{51^{13}}$
3) $\frac{1}{5}$	4) $\left(\frac{1}{5}\right)^4 = \frac{1}{5^4}$
5) $\frac{1}{x^6}$	6) 7
7) $\frac{1}{4x^2}$	8) $9x^2$
9) $\frac{3}{4x}$	10) $\frac{7}{x}$

Name: ....

Powers of Products and Quotients

✓ For any number except zero, a and b and any integer x, $(ab)^x = a^x \times b^x$.

EXAMPLE:

Simplify. $(3x^5y^4)^2 =$

Use Exponent's rules: $(x^a)^b = x^{a \times b}$

$$(3x^5y^4)^2 = (3)^2(x^5)^2(y^4)^2 = 9x^{5 \times 2}y^{4 \times 2} = 9x^{10}y^8$$

PRACTICES:

Simplify.

1) $(5x^3)^2$	2) $(xy)^2$
3) $(ax^2)^3$	4) $(2x^3yz)^2$
5) $(4x^2y^3)^2$	6) $(5x^2y^3)^2$
7) $(2xy^2)^3$	8) $(2x^3y)^4$
9) $(7x^4y^8)^2$	10) $(10x)^3$

Score: ....

Answer Key	
1) $25x^6$	2) x^2y^2
3) a^3x^6	4) $4x^6y^2z^2$
5) $16x^4y^6$	6) $25x^4y^6$
7) $8x^3y^6$	8) $8x^{12}y^4$
9) $49x^8y^{16}$	10) $1,000x^3$

Name: ....

Zero and Negative Exponents

✓ A negative exponent just means that the base is on the wrong side of the fraction line, so you need to flip the base to the other side. For example, "x^{-2}" (pronounced as "ecks to the minus two") just means "x^2" but below, as in $\frac{1}{x^2}$

EXAMPLE:

Evaluate. $\left(\frac{4}{9}\right)^{-2} =$

Use Exponent's rules: $\frac{1}{x^b} = x^{-b} \rightarrow \left(\frac{4}{9}\right)^{-2} = \frac{1}{\left(\frac{4}{9}\right)^2} = \frac{1}{\frac{4^2}{9^2}}$

Now use fraction rule: $\frac{1}{\frac{b}{c}} = \frac{c}{b} \rightarrow \frac{1}{\frac{4^2}{9^2}} = \frac{9^2}{4^2} = \frac{81}{16}$

PRACTICES:

Evaluate the following expressions.

1) 4^{-2}	2) 5^{-2}
3) 6^{-2}	4) 3^{-4}
5) 10^{-1}	6) 33^{-1}
7) 6^{-1}	8) 3^{-2}
9) 9^{-2}	10) 4^{-1}

Score: ....

Answer Key		
1) $\frac{1}{16}$		2) $\frac{1}{25}$
3) $\frac{1}{36}$		4) $\frac{1}{81}$
5) $\frac{1}{10}$		6) $\frac{1}{33}$
7) $\frac{1}{6}$		8) $\frac{1}{9}$
9) $\frac{1}{81}$		10) $\frac{1}{4}$

Name: ....

Negative Exponents and Negative Bases

✓ First make the power positive. A negative exponent can be written as reciprocal of that number with a positive exponent.

✓ The parenthesis is significant!

✓ 5^{-3} is not the same as $(-5)^{-3}$

$(-5)^{-3} = -\dfrac{1}{5^3}$ and $(5)^{-3} = +\dfrac{1}{5^3}$

EXAMPLE:

Simplify. $\left(-\dfrac{5x}{3yz}\right)^{-3} =$

Use Exponent's rules: $\dfrac{1}{x^b} = x^{-b} \rightarrow \left(-\dfrac{5x}{3yz}\right)^{-3} = \dfrac{1}{\left(-\dfrac{5x}{3yz}\right)^3} = \dfrac{1}{-\dfrac{5^3 x^3}{3^3 y^3 z^3}}$

Now use fraction rule: $\dfrac{1}{\frac{b}{c}} = \dfrac{c}{b} \rightarrow \dfrac{1}{\frac{5^3 x^3}{3^3 y^3 z^3}} = -\dfrac{3^3 y^3 z^3}{5^3 x^3} = -\dfrac{27 y^3 z^3}{125 x^3}$

PRACTICES:

Simplify.

1) 7^{-1}	2) $-2x^{-2}$
3) $\dfrac{x}{x^{-5}}$	4) $-\dfrac{a^{-2}}{b^{-1}}$
5) $\dfrac{7}{x^{-5}}$	6) $\dfrac{2b}{-5c^{-2}}$
7) $\dfrac{2n^{-1}}{12p^{-2}}$	8) $\dfrac{8b^{-4}}{3c^{-2}}$
9) $89xy^{-2}$	10) $\left(\dfrac{1}{3}\right)^{-2}$

Score: ....

Answer Key	
1) $\frac{1}{7}$	2) $-\frac{2}{x^2}$
3) x^5	4) $-\frac{b^1}{a^2}$
5) $7x^5$	6) $-2\frac{bc^2}{5}$
7) $\frac{p^2}{6n}$	8) $\frac{8c^2}{3b^4}$
9) $\frac{89x}{y^2}$	10) 9

Name: ....

Writing Scientific Notation

✓ It is used to write very big or very small values in decimal representation.

✓ In scientific notation all numbers can be written in the form of:

$m \times 10^n$, where $1 \leq m \leq 10$ and n is any integer.

Decimal notation	Scientific notation
5	5×10^0
$-25,000$	-2.5×10^4
0.5	5×10^{-1}
2,122.456	2.122456×10^3

EXAMPLE:

Write 8.3×10^{-5} in standard notation.

$10^{-5} \rightarrow$ When the decimal moved to the right, the exponent is negative.

Then: $8.3 \times 10^{-5} = 0.000083$

PRACTICES:

Write each number in scientific notation.

1) 12,000,000	2) 25×10^5
3) 0.0015	4) 54,000
5) 0.0005021	6) 666,012
7) 0.00000076	8) 102,900,000
9) 4,100,000,000	10) 3,600,000

Score: ....

	Answer Key	
1) 1.2×10^7		2) 2.5×10^6
3) 1.5×10^{-3}		4) 5.4×10^4
5) 5.021×10^{-4}		6) 6.66012×10^5
7) 7.6×10^{-7}		8) 1.029×10^8
9) 4.1×10^9		10) 3.6×10^6

Name: ....

Square Roots

✓ A square root of x is a number p whose square is: $p^2 = x$

p is a square root of x.

EXAMPLE:

Find the square root of $\sqrt{225}$.

First factor the number: $225 = 15^2$, Then: $\sqrt{225} = \sqrt{15^2}$

Now use radical rule: $\sqrt[n]{a^n} = a$

Then: $\sqrt{15^2} = 15$

PRACTICES:

Find the value each square root.

1) $\sqrt{25}$	2) $\sqrt{1,600}$
3) $\sqrt{100}$	4) $\sqrt{121}$
5) $\sqrt{4}$	6) $\sqrt{225}$
7) $\sqrt{10,000}$	8) $\sqrt{16}$
9) $\sqrt{64}$	10) $\sqrt{36}$

Score: ....

Answer Key	
1) 5	2) 40
3) 10	4) 11
5) 2	6) 15
7) 100	8) 4
9) 8	10) 6

Chapter 5 : Algebraic Expressions

Topics that you'll learn in this chapter:

- ➢ Expressions and Variables

- ➢ Simplifying Variable and Polynomial Expressions

- ➢ Translate Phrases into an Algebraic Statement

- ➢ The Distributive Property

- ➢ Evaluating One and two Variable

- ➢ Combining like Terms

"Without mathematics, there's nothing you can do. Everything around you are mathematics. Everything around you are numbers." – Shakuntala Devi

Name: ....

Translate Phrases into an Algebraic Statement

How to translate key words and phrases into algebraic expressions:

- ✓ Addition: the sum of, more than, plus, etc.
- ✓ Subtraction: less than, decreased, minus, etc.
- ✓ Multiplication: times, multiplied, product, etc.
- ✓ Division: quotient, ratio, divided, etc.

EXAMPLE:

5 times the sum of x and 8

Sum of 8 and x: $8 + x$. Times is used for multiplication. Then: $5 \times (8 + x)$

PRACTICES:

Write an algebraic expression for each phrase.

1) Fifteen subtracted from a number.

2) The quotient of seventeen and a number.

3) A number increased by fifty.

4) A number divided by – 21.

5) The difference between sixty –three and a number.

6) Threefold a number decreased by 45.

7) Seven times the sum of a number and – 21.

8) The quotient of 90 and the product of a number and – 8.

9) Nine subtracted from 4 times a number.

10) The difference of six and a number.

Score: ....

Answer Key	
1) $x - 15$	2) $\dfrac{17}{x}$
3) $x + 50$	4) $-\dfrac{x}{21}$
5) $63 - x$	6) $3x - 45$
7) $7(x + (-21))$	8) $-\dfrac{90}{8x}$
9) $4x - 9$	10) $6 - x$

Name: ....

The Distributive Property

✓ Distributive Property:

$$a(b + c) = ab + ac$$

EXAMPLE:

Simply. $(5x - 3)(-5) =$

Use Distributive Property formula: $a(b + c) = ab + ac$

$$(5x - 3)(-5) = -25x + 15$$

PRACTICES:

Use the distributive property to simply each expression.

1) $4(9 - 3x)$	2) $-(-8 - 4x)$
3) $(-5x - 1)(-2)$	4) $(-3)(2x - 4)$
5) $4(5 + 3x)$	6) $(-9x + 10)3$
7) $(-4 - 5x)(-3)$	8) $(-2x)(-3 + 2x) - 3x(1 - 5x)$
9) $(-2)(3x - 1) + 4(3x + 2)$	10) $(-15)(2x + 3)$

Score: ....

Answer Key

1) $-12x + 36$	2) $4x + 8$
3) $10x + 2$	4) $-6x + 12$
5) $12x + 20$	6) $-27x + 30$
7) $15x + 12$	8) $11x^2 + 3x$
9) $6x + 10$	10) $-30x - 45$

Name: ....

Evaluating One Variable

✓ To solve a variable expression, find the variable and substitute a number for that variable.

✓ Perform the mathematical operations.

EXAMPLE:

Solve this expression. $12 - 2x, x = -1$

First substitute -1 for x, then:

$$12 - 2x = 12 - 2(-1) = 12 + 2 = 14$$

PRACTICES:

Simplify each algebraic expression.

1) $5x + 4, x = 1$	2) $x + (-4), x = -6$
3) $-10x + 8, x = -2$	4) $\left(-\frac{36}{x}\right) - 10 + 2x, x = 6$
5) $\frac{36}{x} - 3, x = 3$	6) $(-10) - \frac{x}{4} + 4x, x = -8$
7) $15 + 6x - 3, x = -1$	8) $(-5) + \frac{x}{8}, x = 64$
9) $\left(-\frac{24}{x}\right) - 10 + 5x, x = 4$	10) $(-4) + \frac{4x}{9}, x = 81$

Score: ....

Answer Key	
1) 9	2) −10
3) 28	4) −4
5) 9	6) −40
7) 6	8) 3
9) 4	10) 32

Name: ....

Evaluating Two Variables

To solve an algebraic expression, substitute a number for each variable and perform the mathematical operations.

EXAMPLE:

Solve this expression. $-3x + 5y, x = 2, y = -1$

First substitute 2 for x, and -1 for y, then:

$$-3x + 5y = -3(2) + 5(-1) = -6 - 5 = -11$$

PRACTICES:

Simplify each algebraic expression.

1) $5a - (5 - b)$,

$a = 2, b = 3$

2) $5x + 3y - 6 + 3y$,

$x = 3, y = 1$

3) $\left(-\frac{27}{x}\right) + 4 + 3y$,

$x = 3, y = 5$

4) $(-4)(-3a - 5b)$,

$a = 3, b = 4$

5) $7x + 10 - 5y$,

$x = 3, y = 6$

6) $18 + 3(-x - 4y)$,

$x = 2, y = 5$

7) $12x + 2y$,

$x = 5, y = 10$

8) $x \times 6 \div 3y$,

$x = 6, y = 1$

9) $4x - 3y$,

$x = 6, y = 3$

10) $\left(-\frac{14}{x}\right) + 4y$,

$x = 7, y = -3$

Score: ....

Answer Key	
1) 8	2) 15
3) 10	4) 116
5) 1	6) − 48
7) 80	8) 12
9) 15	10) −14

Name: ....

Expressions and Variables

- ✓ In algebra, a variable is a letter used as a replacement for a number. The most common letters are: $x, y, z, a, b, c, m,$ and n.
- ✓ An algebraic expression is an expression that has variables, integers, and math operations such as addition, subtraction, multiplication, division, etc.
- ✓ In an expression, we can combine "identical" terms. (Values with same power and variable)

EXAMPLE:

Simplify this expression. $(10x + 2x + 3) = ?$

Combine like terms. Then: $(10x + 2x + 3) = 12x + 3$ (remember you cannot combine variables and numbers

PRACTICES:

Simplify each expression.

1) $10(-3 - 8x), x = 4$	2) $-3(5 - 8x) - 6x, x = 1$
3) $2x - 8x, x = 2$	4) $x + 12x, x = 6$
5) $20 - 5x + 10x + 5, x = 3$	6) $15(5x + 3), x = 0$
7) $20(4 - x) - 9, x = 2$	8) $20x - 8x - 10, x = 5$
9) $6x + 9y, x = 4, y = 2$	10) $6x - 2x, x = 8$

Score: ....

Answer Key	
1) -350	2) 3
3) -12	4) 78
5) 40	6) 45
7) 31	8) 50
9) 42	10) 32

Name: ....

Combining like Terms

✓ We separate the terms by "+" and "-" signs.
✓ Identical terms are those terms with same powers and same variables.
Make sure to use the "+" or "-" that is in front of the coefficient

EXAMPLE:

Simplify this expression. $(-5)(8x - 6) =$

Use Distributive Property formula: $a(b + c) = a + ac$

$(-5)(8x - 6) = -40x + 30$

PRACTICES:

Simplify each expression.

1) $-8(-5x + 1)$	2) $6(-2 + 4x)$
3) $-8 - 14x + 16x + 3$	4) $9x - 7x - 15 + 18$
5) $(-9)(12x - 21) + 31$	6) $2(4x + 9) + 12x$
7) $4(-2x - 17) + 14(3x + 1)$	8) $(9x - 5y)7 + 25y$
9) $4.5x^3 \times (-8x)$	10) $-19 - 15x^2 + 12x^2$

Score: ....

Answer Key

1) $40x - 8$	2) $24x - 12$
3) $2x - 5$	4) $2x + 3$
5) $220 - 108x$	6) $20x + 18$
7) $34x - 54$	8) $63x - 10y$
9) $-36x^4$	10) $-3x^2 - 19$

Name: ....

Simplifying Polynomial Expressions

✓ A polynomial is a unique expression that consists of coefficients and variables that performs only the arithmetic of addition, subtraction, multiplication, and non-negative integer exponents of variables.

$$P(x) = a_n x^n + a_{n-1} x^{n-1} + \ldots + a_2 x^2 + a_1 x + a_0$$

EXAMPLE:

Simplify this Polynomial Expressions. $4x^2 - 5x^3 + 15x^4 - 12x^3 =$

Combine "like" terms: $-5x^3 - 12x^3 = -17x^3$

Then: $4x^2 - 5x^3 + 15x^4 - 12x^3 = 4x^2 - 17x^3 + 15x^4$

Then write in standard form: $4x^2 - 17x^3 + 15x^4 = 15x^4 - 17x^3 + 4x^2$

PRACTICES:

Simplify each polynomial.

1) $(2x^2 + 4) - (9 + 5x^2)$	2) $(25x^3 - 12x^2) - (6x^2 - 9x^3)$
3) $14x^5 - 15x^6 + 2x^5 - 16x^6 + x^6$	4) $(15 + 12x^3) + (3x^3 + 5)$
5) $13x^3 - 15x^4 + 12x^3 + 20x^4$	6) $-6x^2 + 15x^2 + 17x^3 + 16 - 32$
7) $15x^3 + 12 + 2x^2 - 5x - 10x$	8) $24x^2 - 16x^3 - 4x(2x^2 + 3x)$
9) $(21x^4 - 10x) - (2x - x^4)$	10) $(7x^2 - 9) + (x^2 - 8x^3)$

Score: ....

Answer Key	
1) $-3x^2 - 5$	2) $34x^3 - 18x^2$
3) $-30x^6 + 16x^5$	4) $15x^3 + 20$
5) $5x^4 + 25x^3$	6) $17x^3 + 9x^2 - 16$
7) $15x^3 + 2x^2 - 15x + 12$	8) $-24x^3 + 12x^2$
9) $22x^4 - 12x$	10) $-8x^3 + 8x^2 - 9$

Chapter 6 : Equations and Inequalities

Topics that you'll learn in this chapter:

➤ One, Two, and Multi – Step Equations

➤ Graphing Single– Variable Inequalities

➤ One, Two, and Multi – Step Inequalities

➤ Solving Systems of Equations by Substitution and Elimination

➤ Finding Slope and Writing Linear Equations

➤ Graphing Lines Using Slope– Intercept and Standard Form

➤ Graphing Linear Inequalities

➤ Finding Midpoint and Distance of Two Points

"The study of mathematics, like the Nile, begins in minuteness but ends in magnificence." – *Charles Caleb Colton*

Name: ....

One–Step Equations

✓ The values of two algebraic expressions on both sides of an equation are always equal.

$$ax + b = c$$

✓ You only require performing one Math operation to solve the problem.

EXAMPLE:

Solve this equation. $x + 24 = 0 , x = ?$

Here, we have the addition operation, and its inverse operation is subtraction. To solve this equation, subtract 24 from both sides of the equation: $x + 24 - 24 = 0 - 24$

Then simplify: $x + 24 - 24 = 0 - 24 \rightarrow x = -24$

PRACTICES:

Solve each equation.

1) $x + 4 = 16$	2) $48 = (-2) + x$
3) $5x = (-105)$	4) $(-8) = (8x)$
5) $(-2) = 14 + x$	6) $5 + x = 6$
7) $2x + 3 = (-7)$	8) $28 = x + 7$
9) $(-15) + x = (-15)$	10) $12x = (-36)$

Score: ....

Answer Key

1) 12	2) 50
3) −21	4) −1
5) −16	6) 1
7) −5	8) 21
9) 0	10) −3

Name: ....

Two–Step Equations

- ✓ You only require performing two math operations (add, subtract, multiply, or divide) to solve the equation.
- ✓ Simplify the equation using the inverse of addition or subtraction.
- ✓ Simplify the equation further by using the inverse of division or multiplication.

EXAMPLE:

Solve this equation. $3x = 15, x =?$

Here, we have the multiplication operation (variable x is multiplied by 3) and its inverse operation is division. To solve this problem, we divide both sides of equation by 3:

$3x = 15 \rightarrow 3x \div 3 = 15 \div 3 \rightarrow x = 5$

PRACTICES:

Solve each equation.

1) $4(2 + 2x) = 8$	2) $(-5)(x - 3) = 25$
3) $(-5)(2x - 5) = (-15)$	4) $4(9 + 3x) = -12$
5) $6(2x + 1) = 30$	6) $2(x + 2) = 42$
7) $2(12 + 6x) = 60$	8) $(-10)(5x) = 100$
9) $4(3x + 3) = 24$	10) $\dfrac{x - 5}{3} = 4$

Score: ....

Answer Key	
1) 0	2) −2
3) 4	4) −4
5) 2	6) 10
7) 3	8) −2
9) 1	10) 17

Name: ....

Multi–Step Equations

✓ Combine "identical" terms on one side.

✓ Put variables to one side by adding or subtracting.

✓ Simplify the equation by using the inverse of addition or subtraction.

✓ Simplify further by using the inverse of division or multiplication.

EXAMPLE:

Solve this equation. $-(2 - x) = 5$

First, use Distributive Property: $-(2 - x) = -2 + x$

Now by adding 2 to both sides of the equation, we can solve it.

$-2 + x = 5 \rightarrow -2 + x + 2 = 5 + 2$
Now simplify: $-2 + x + 2 = 5 + 2 \rightarrow x = 7$

PRACTICES:

Solve each equation.

1) $8 - 2x = 28$	2) $-10 = -(x + 7)$
3) $2x - 17 = (-x) + 1$	4) $-2x = (-3x) - 8$
5) $5(14 + 2x) + 3x = -x$	6) $x - 11 = x - 5 + 2x$
7) $15 + 2x = (-25) - 2x + 3x$	8) $-3(x - 3x) = 40 - 4x$
9) $24 + 8x + x = (-x + 4)$	10) $-8(1 + 5x) = 152$

Score: ....

Answer Key	
1) -10	2) 3
3) 6	4) -8
5) -5	6) -3
7) -40	8) 4
9) -2	10) -4

Name: ....

Graphing Single–Variable Inequalities

- ✓ Inequality is like equations and uses symbols for "less than" (<) and "greater than" (>).
- ✓ To solve inequalities, we need to separate the variable. (Like in equations)
- ✓ Find the value of the inequality on the number line to graph an inequality.
- ✓ For greater than or less than draw open circle on the value of the variable.
- ✓ Use filled circle if there is an equal sign too.
- ✓ Draw a line to the left or to the right for less or greater than.

EXAMPLE:

Draw a graph for $x > 2$

Since, the variable is greater than 2, then we need to find 2 and draw an open circle above it. Then, draw a line to the right.

PRACTICES:

Draw a graph for each inequality.

1) $2 \geq x$

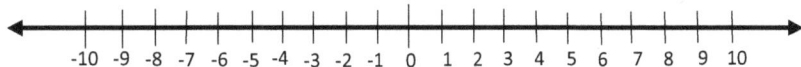

2) $x < 3$

3) $5 \geq x$

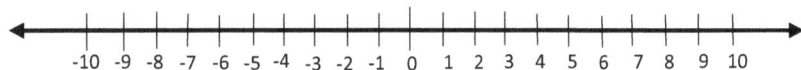

4) $x \geq -2$

5) $x > 0$

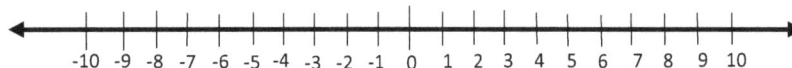

6) $-1.5 < x$

7) $x \geq -1$

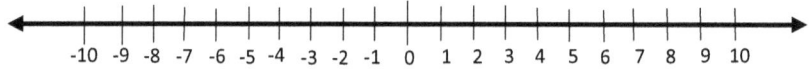

Score: ....

Answer Key

1) $2 \geq x$

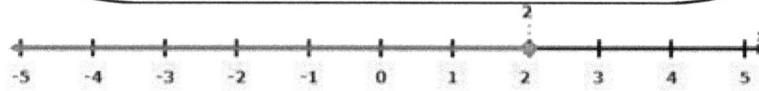

2) $x < 3$

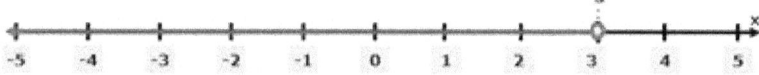

3) $5 \geq x$

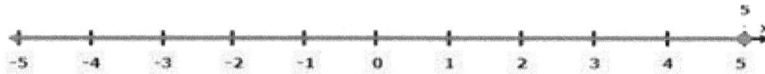

4) $x \geq -2$

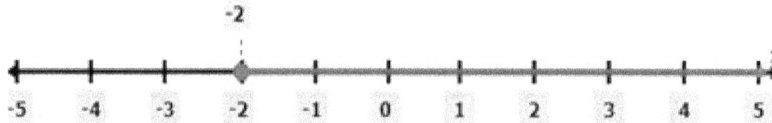

5) $x > 0$

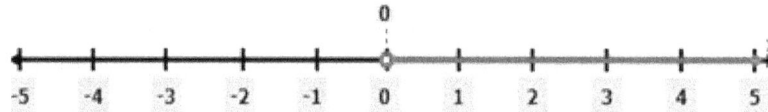

6) $-1.5 < x$

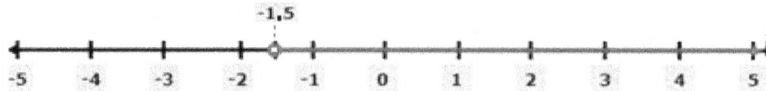

7) $x \geq -1$

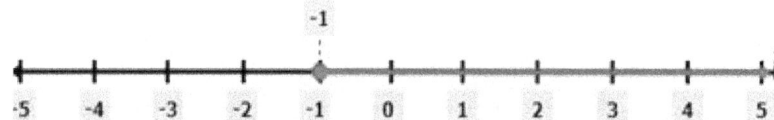

Name: ....

One–Step Inequalities

✓ Like equations, first separate the variable by using inverse operation.
✓ For dividing or multiplying both sides by negative numbers, flip the direction of the inequality sign.

EXAMPLE:

Solve this inequality. $x - 1 \leq 2$

Add 1 to both sides. $x - 1 \leq 2 \rightarrow x - 1 + 1 \leq 2 + 1$, then: $x \leq 3$

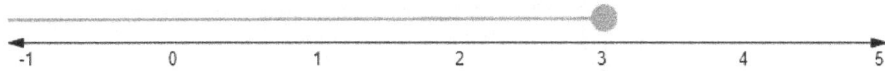

PRACTICES:

Solve each inequality and graph it.

1) $2x + 3 \geq 7$

2) $x < 3$

3) $5 \geq x$

4) $x \geq -2$

5) $x > 0$
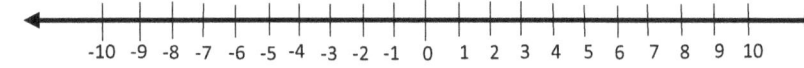

6) $-1.5 < x$
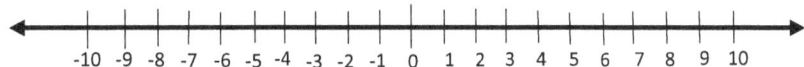

7) $x \geq -1$
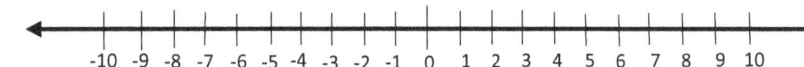

Score: ....

Answer Key

1) $2 \geq x$

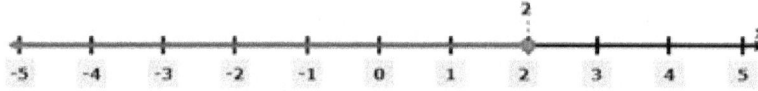

2) $x < 3$

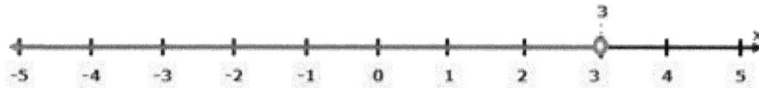

3) $5 \geq x$

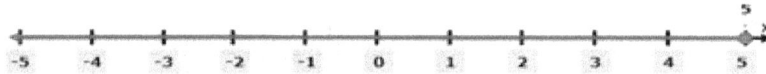

4) $x \geq -2$

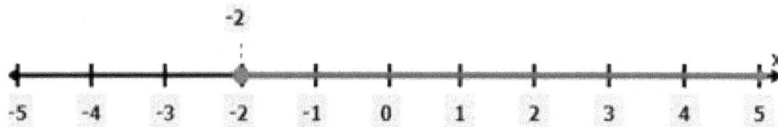

5) $x > 0$

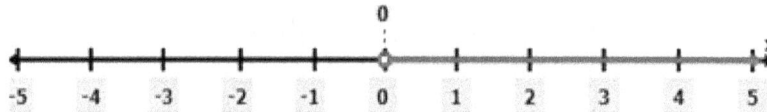

6) $-1.5 < x$

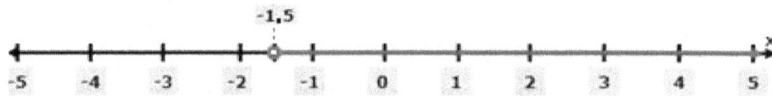

7) $x \geq -1$

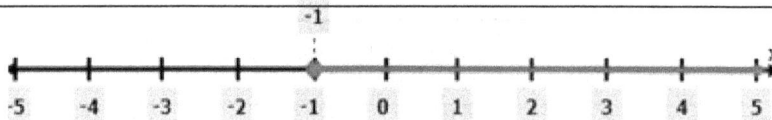

Name: ....

Two–Step Inequalities

✓ Separate the variable.

✓ Flip the direction of the inequality sign for dividing both sides by negatives numbers.

✓ We can simplify by using the inverse of addition or subtraction.

✓ We can simplify further by using the inverse of division or multiplication.

EXAMPLE:

Solve: $2x + 9 \geq 11$

First add -9 to both sides: $2x + 9 - 9 \geq 11 - 9 \rightarrow 2x \geq 2$

Now, divide both sides by 2: $2x \geq 2 \rightarrow x \geq 1$

PRACTICES:

Solve each inequality and graph it.

1) $x - 4 \leq 4$	2) $x + 4 \geq 5$
3) $3x - 2 \leq 7$	4) $5x + 2 < 12$
5) $x + 7 \geq 9$	6) $3x - 3 \leq 3$
7) $7x - 4 < 3$	8) $8 + x \leq 13$
9) $2x + 7 \leq 11$	10) $10x - 16 < 4$

Score: ....

Answer Key	
1) $x \leq 8$	2) $x \geq 1$
3) $x \leq 3$	4) $x < 2$
5) $x \geq 2$	6) $x \leq 2$
7) $x < 1$	8) $x \leq 5$
9) $x \leq 2$	10) $x < 2$

Name: ....

Multi–Step Inequalities

✓ Separate the variable.

✓ We can simplify by using the inverse of addition or subtraction.

✓ We can simplify further by using the inverse of division or multiplication.

EXAMPLE:

Solve this inequality. $2x - 2 \leq 6$

First add 2 to both sides: $2x - 2 + 2 \leq 6 + 2 \rightarrow 2x \leq 8$

Now, divide both sides by 2: $2x \leq 8 \rightarrow x \leq 4$

PRACTICES:

Solve each inequality.

1) $-(x + 3) + 8 < 25$	2) $\frac{3x + 1}{2} \leq 5$
3) $\frac{x - 4}{3} > 7$	4) $4(x - 2) \leq 8$
5) $\frac{x}{3} + \frac{1}{3} < 2$	6) $\frac{x+4}{5} > 3$
7) $\frac{x}{8} + \frac{3}{4} < 1$	8) $2(x + 5) + 4 > 10$
9) $24 + 5x < 4$	10) $\frac{x+2}{3} > 10$

Score: ....

Answer Key	
1) $x > -20$	2) $x \leq 3$
3) $x > 25$	4) $x \leq 4$
5) $x < 5$	6) $x > 11$
7) $x < 2$	8) $x > -2$
9) $x < -4$	10) $x > 28$

Name: ....

Solving Systems of Equations by Substitution

✓ Let the system of equations. $x + y = 1$; $-2x + y = 4$

Put $x = 1 - y$ in the second equation.

$-2(1 - y) + y = 4 \rightarrow -2 + 2y + y = 4 \Rightarrow y = 2$

Put $y = 2$ in $x = 1 - y$; then $x = 1 - 2 = -1$; $(-1, 2)$

EXAMPLE:

Solve: $-2x - 2y = -13$; $-4x + 2y = 10$

For the first equation above, you can add $-4x + 2y$ to the left side and 10 to the right side of the first equation: $-2x - 2y + (-4x + 2y) = -13 + 10$. Now, if you simplify, you get: $-2x - 2y - 4x + 2y = -3 \rightarrow -6x = -3 \rightarrow$ $x = 0.5$. Now, put 0.5 for the x in the first equation: $-2(0.5) - 2y = -13$. By solving this equation, $y = 6$

PRACTICES:

Solve each system of equation by substitution.

1) $-x + 5y = -4$ $\quad x - 3y = 8$	2) $2x + 3y = -6$ $\quad -2x - y = 8$
3) $x + 2y = -5$ $\quad 5x - 10y = 5$	4) $y = -x + 5$ $\quad 3x - y = -3$
5) $3x = 6$ $\quad 10y = 4x + 2$	6) $3x + 2y = 2$ $\quad x + 4y = -6$
7) $4x + y = 3$ $\quad 2x + 4y = -2$	8) $4y = 2x + 3$ $\quad x - 4y = -2$
9) $7y = 14x$ $\quad 2x - 5y = -24$	10) $5y = x + 2$ $\quad 3x - 12y = -5$

Score: ....

Answer Key	
1) $(14, 2)$	2) $(-\frac{9}{2}, 1)$
3) $(-2, -\frac{3}{2})$	4) $(\frac{1}{2}, \frac{9}{2})$
5) $(2, 1)$	6) $(2, -2)$
7) $(1, -1)$	8) $(-1, \frac{1}{4})$
9) $(3, 6)$	10) $(-\frac{1}{3}, \frac{1}{3})$

Name: ....

Solving Systems of Equations by Elimination

✓ A system of equations has two equations and two variables. For example, look at the system of equations: $x - y = 1, x + y = 5$

✓ The simplest way to solve a system of equation is using the elimination method. The elimination method uses the addition property of equality. On each side of an equation, you can add the same value.

EXAMPLE:

What is the value of x *and* y in this system of equations? $\begin{cases} 3x - 4y = -20 \\ -x + 2y = 10 \end{cases}$

Solving Systems of Equations by Elimination: $\begin{array}{l} 3x - 4y = -20 \\ -x + 2y = 10 \end{array}$ ⇒ Multiply the second

equation by 3, then add it to the first equation.

$\begin{array}{l} 3x - 4y = -20 \\ 3(-x + 2y = 10) \end{array} \Rightarrow \begin{array}{l} 3x - 4y = -20 \\ -3x + 6y = 30) \end{array} \Rightarrow 2y = 10 \Rightarrow y = 5.$ Now, substitute 5 for y in

the first equation and solve for x. $3x - 4(5) = -20 \rightarrow 3x - 20 = -20 \rightarrow x = 0$

PRACTICES:

Solve each system of equation by elimination.

1) $-5x + y = -5$
   $-y = -6x + 6$

2) $-6x - 2y = -2$
   $2x - 3y = 8$

3) $5x - 4y = 8$
   $-6x + y = -21$

4) $10x - 4y = -24$
   $-x - 20y = -18$

5) $25x + 3y = -13$
   $12x - 6y = -36$

6) $x - 8y = -7$
   $6x + 4y = 10$

7) $-6x + 16y = 4$
   $5x + y = 11$

8) $2x - 3y = -10$
   $4x + 6y = -20$

9) $x - 5y = -8$
   $3x + 7y = -2$

10) $x - 2y = -3$
    $2x + 6y = -1$

Score: ....

	Answer Key	
1) $(1,0)$		2) $(1,-2)$
3) $(4,3)$		4) $(-2,1)$
5) $(-1,4)$		6) $(1,1)$
7) $(2,1)$		8) $(-5,0)$
9) $(-3,1)$		10) $(-2,\frac{1}{2})$

Name: ....

Systems of Equations Word Problems

✓ Define your variables, write the system of equations then use elimination method for solving systems of equations.

EXAMPLE:

Tickets to a movie cost $5 for students and $8 for adults. Some friends purchased **20 tickets** for **$115.00.** How many adults ticket did they buy?

Let x be the number of adult tickets and y be the number of student tickets. There are 20 tickets. Then: $x + y = 20$. The cost of students' ticket is $5 and for adults it is $8, and the total cost is $115. So, $8x + 5y = 20$. Now, we have a system of equations:

$$\begin{cases} x + y = 20 \\ 8x + 5y = 115 \end{cases}$$

Multiply the first equation by -5 and add to the second equation: $-5(x + y = 20) =$

$-5x - 5y = -100$

$8x + 5y + (-5x - 5y) = 115 - 100 \rightarrow 3x = 15 \rightarrow x = 5 \rightarrow 5 + y = 20 \rightarrow y = 15$.

There are 5 adults' tickets and 15 student tickets.

PRACTICES:

Solve.

1) A school of 220 students went on a field trip. They took 20 vehicles, some vans, and some minibuses. Find the number of vans and the number of minibuses they took if each van holds 5 students and each minibus hold 15 students.

2) The sum of two numbers is 28. Their difference is 12. Find the numbers.

3) A farmhouse shelters 20 animals, some are pigs, and some are gooses. Altogether there are 64 legs. How many of each animal are there?

4) The sum of the digits of a certain two–digit number is 15. Reversing it's increasing the number by 9. What is the number?

5) The difference of two numbers is 16. Their sum is 32. Find the numbers.

6) Tickets to a movie cost $5 for adults and $3 for students. A group of friends purchased 18 tickets for $82.00. How many adults ticket did they buy?

7) At a store, Eva bought two shirts and five hats for $154.00. Nicole bought three same shirts and four same hats for $168.00. What is the price of each shirt?

8) A farmhouse shelters 10 animals, some are pigs, and some are ducks. Altogether there are 36 legs. How many pigs are there?

9) A class of 195 students went on a field trip. They took 19 vehicles, some cars and some buses. If each car holds 5 students and each bus hold 25 students, how many buses did they take?

10) The sum of two numbers is 50. Their difference is 20. Find the numbers.

Score: ....

Answer Key	
1) There are 8 van and 12 minibuses.	2) 8 and 20
3) There are 12 pigs and 8 gooses.	4) 78
5) 24 and 8.	6) 14
7) $32	8) 8
9) 5	10) 15 and 35

Linear Equations

✓ The equation of a line: $y = mx + b$

✓ Find the slope.

✓ Find the y-intercept. This can be done by substituting the slope and the coordinates of a point (x, y) on the line.

EXAMPLE:

What will be the equation of the line that passes through $(2, -2)$ and has a slope of 7?

$y = mx + b$ is the general slope-intercept form of the equation; where, m is the slope and b is the y-intercept.

By substitution of the given point and given slope, we have: $-2 = (2)(7) + b$

So, $b = -2 - 14 = -16$, and our required equation is $y = 7x - 16$.

PRACTICES:

Find the slope of the line through each pair of points.	Write the slope–intercept form of the equation of the line through the given points.
1) $(3, 1), (2, 4)$	2) Through: $(2,3), (4,2)$
3) $(-3, 4), (-1, 6)$	4) Through: $(8, -3), (6, 7)$
5) $(4, 4), (6, -6)$	6) Through: $(0.5, 4), (2.5, 4.4)$
7) $(-1, 8), (5, -4)$	8) Through: $(4, -2), (2.5, 1)$
9) $(12, -3), (7, -3)$	10) Through: $(-1, 0.7), (-2.3, 2)$

Score: ....

Answer Key	
1) −3	2) $y = -\frac{1}{2}x + 4$
3) 1	4) $y = -5x + 37$
5) − 5	6) $y = \frac{1}{5}x + \frac{39}{10}$
7) −2	8) $y = -2x + 6$
9) 0	10) $y = -x - 0.3$

Name: ....

Graphing Lines of Equations

✓ The equation of the line is: $y = mx + c$

If slope-intercept form of a line given the slope m and the y-intercept (the intersection of the line and y-axis)

EXAMPLE:

Sketch the graph of $y = 8x - 3$.

To graph this line, two points are required.

We have value of y is -3 when x is 0.

The value of x is 3/8 when y is 0.

$x = 0 \rightarrow y = 8(0) - 3 = -3,$

$y = 0 \rightarrow 0 = 8x - 3 \rightarrow x = \dfrac{3}{8}$

Now, the two points are: $(0,-3)$ and $(\frac{3}{8},0)$.

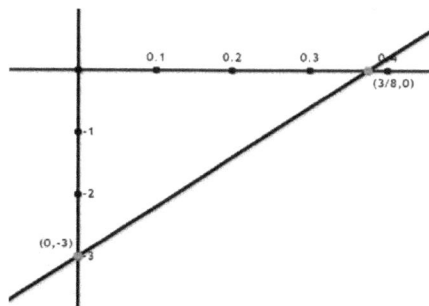

Find the points and graph the line. The slope of the line is 8.

PRACTICES:

Sketch the graph of each line.

1) $y = 3x - 2$

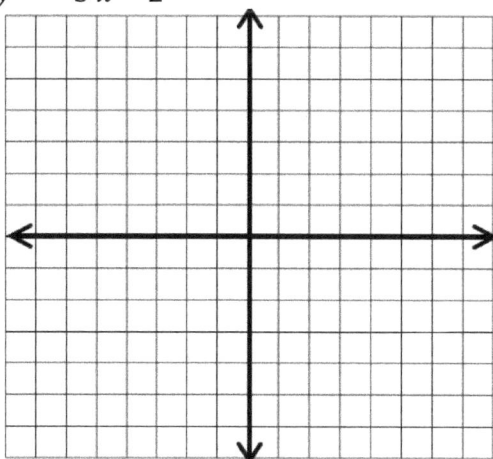

2) $y = 2x + 3$

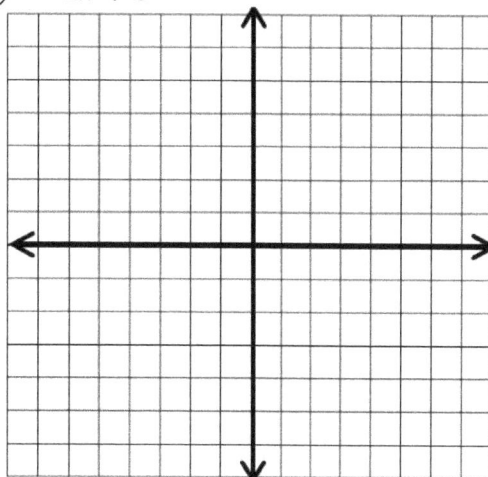

3) $-2x = y + 5$

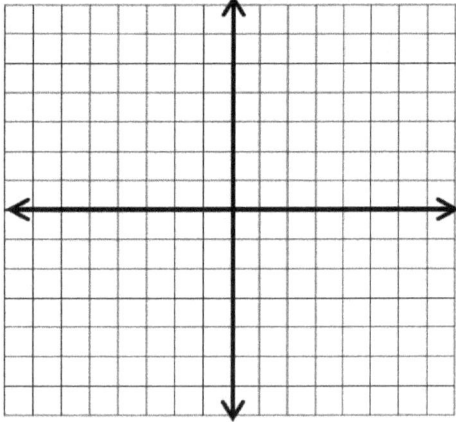

4) $4x + y = 2$

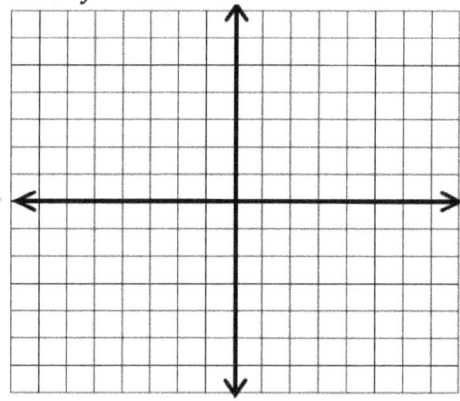

Score: ....

Answer Key

1)

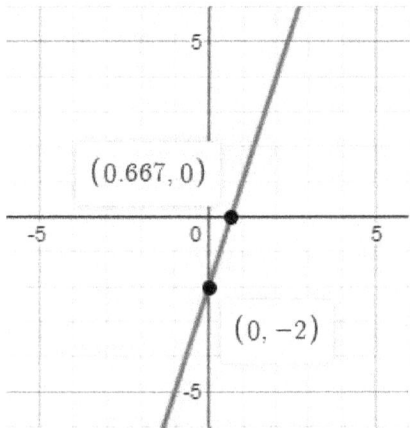

$(0.667, 0)$

$(0, -2)$

2)

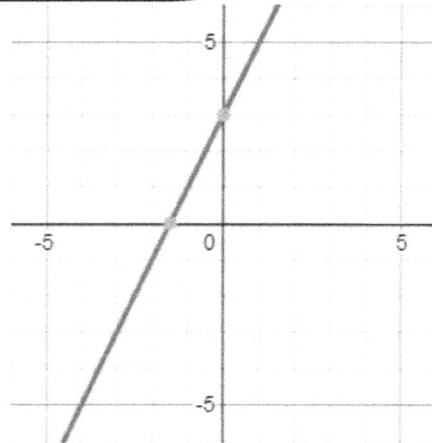

3)

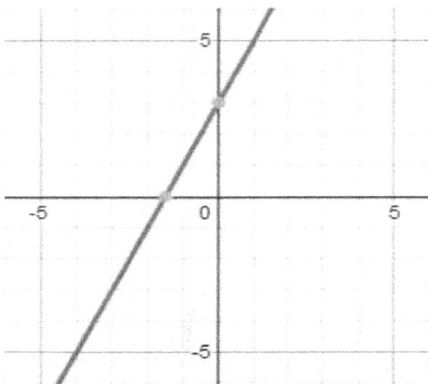

4)

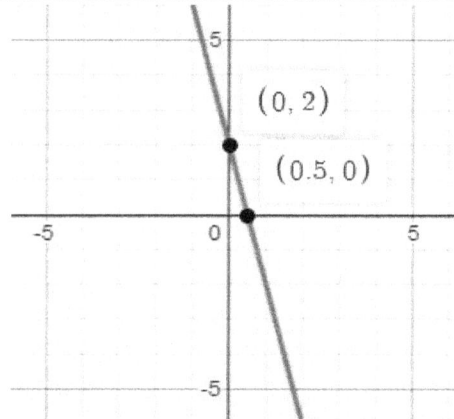

$(0, 2)$

$(0.5, 0)$

Name: ....

Graphing Linear Inequalities

✓ First step is to graph the "equals" line.
✓ Choose a testing point. (It can be any point on both sides of the line.)
✓ Put the value of (x, y) of that point in the inequality. If this satisfy the inequality, then this part of line is solution. If it does not satisfy then other part of line, is solution.

EXAMPLE:

Plot the graph of $y < 2x - 3$.

We know that first step is to graph the line: $y = 2x - 3$.

Now y-intercept is -3 and slope is 2. Then, select a testing point. The simplest point to test is the origin:$(0, 0)$

$(0,0) \rightarrow y < 2x - 3 \rightarrow 0 < 2(0) - 3 \rightarrow 0 < -3$

0 is greater than -3. So, the other part of the line

(On the right side) is the solution.

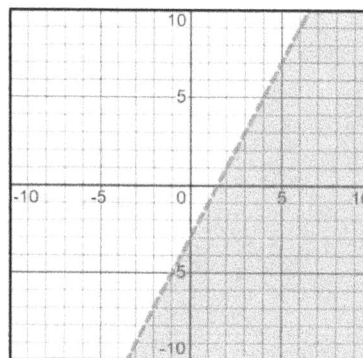

PRACTICES:

Sketch the graph of each linear inequality.

1) $2y + 8x \geq 4$

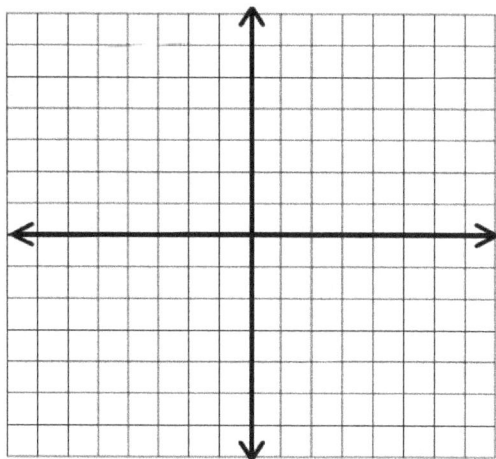

2) $-x + 2y \leq 4$

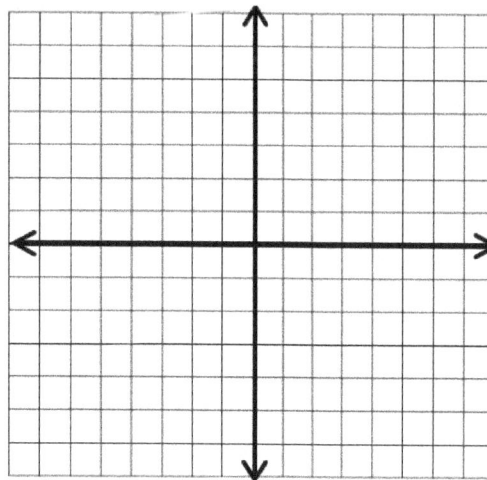

3) $2x + \frac{1}{2}y < 2$

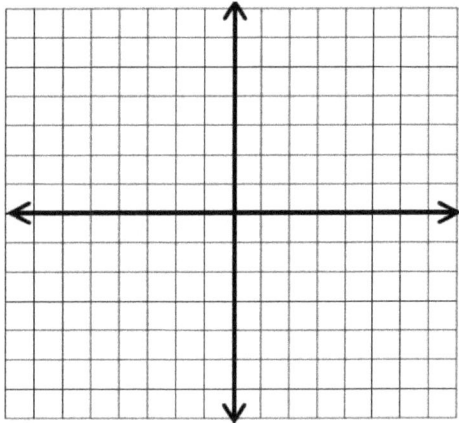

4) $-\frac{1}{3}x + y < 2$

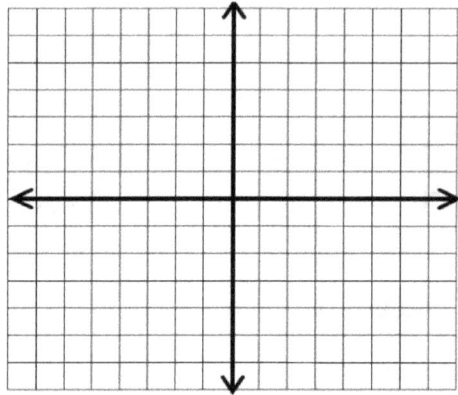

Score: ....

Answer Key

1)

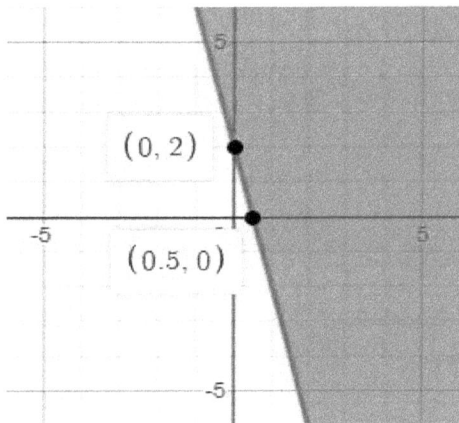

$(0, 2)$

$(0.5, 0)$

2)

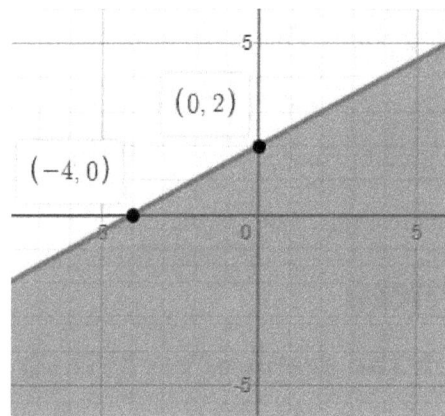

$(0, 2)$

$(-4, 0)$

3)

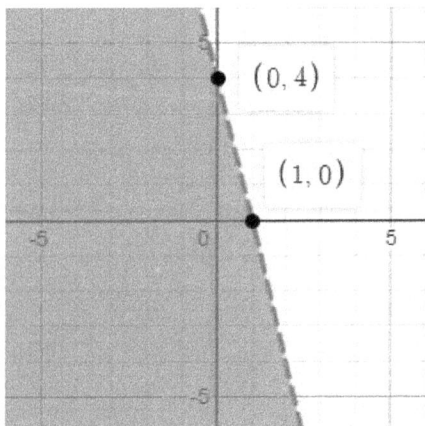

$(0, 4)$

$(1, 0)$

4)

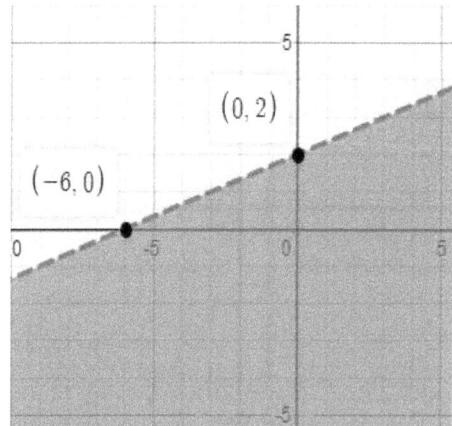

$(0, 2)$

$(-6, 0)$

Name: ....

Finding Distance of Two Points

✓ Distance of wo points A (x_1, y_1) and B (x_2, y_2): $d = \sqrt{(x_1 - x_2)^2 + (y_1 - y_2)^2}$

✓ The midpoint is middle of a line segment.

✓ We can find Midpoint of two endpoints A (x_1, y_1) and B (x_2, y_2) using this formula: $M\left(\frac{x_1+x_2}{2}, \frac{y_1+y_2}{2}\right)$

EXAMPLE:

Find the distance between of $(0, 8), (-4, 5)$.

Use distance of two points formula: $d = \sqrt{(x_1 - x_2)^2 + (y_1 - y_2)^2}$

$(x_1, y_1) = (0, 8)$ and $(x_2, y_2) = (-4, 5)$. Then: $d = \sqrt{(x_1 - x_2)^2 + (y_1 - y_2)^2} \rightarrow$

$d = \sqrt{(0 - (-4))^2 + (8 - 5)^2} = \sqrt{(4)^2 + (3)^2} = \sqrt{16 + 9} = \sqrt{25} = 5 \rightarrow d = 5$

PRACTICES:

Find the midpoint of the line segment with the given endpoints.	Find the distance between each pair of points.
1) $(1.5, -1), (0.5, -1)$	2) $(3, 4), (2, -1)$
3) $(1.5, -1), (0.5, -1)$	4) $(6, -1), (2,3)$
5) $(0, 3), (4, -9)$	6) $(2, 5), (-2, 5)$
7) $(5, 2), (1, 5)$	8) $(0, -4), (-5, 1)$
9) $(-2, 0), (3, -4)$	10) $(3, -2), (-1, -5)$

Score: ....

Answer Key	
1) $(1, -1)$	2) 5.09
3) $(1.5, -2)$	4) 5.656
5) $(2, -3)$	6) 4
7) $(3, 3.5)$	8) 7.07
9) $(0.5, -2)$	10) 5

Chapter 7 : Polynomials

Topics that you'll learn in this chapter:

- ➢ Classifying Polynomials

- ➢ Writing Polynomials in Standard Form

- ➢ Simplifying Polynomials

- ➢ Adding and Subtracting Polynomials

- ➢ Multiplying and Dividing Monomials

- ➢ Multiplying a Polynomial and a Monomial

- ➢ Multiplying Binomials

- ➢ Factoring Trinomials

- ➢ Operations with Polynomials

"Mathematics – the unshaken Foundation of Sciences, and the plentiful Fountain of Advantage to human affairs." — *Isaac Barrow*

Name: ....

Classifying Polynomials

Name	Degree	Example
constant	0	4
linear	1	$2x$
quadratic	2	$x^2 + 5x + 6$
cubic	3	$x^3 - x^2 + 4x + 8$
quartic	4	$x^4 + 3x^3 - x^2 + 2x + 6$
quantic	5	$x^5 - 2x^4 + x^3 - x^2 + x + 10$

EXAMPLE:

$17x^5 \Rightarrow$ Quantic binomial

PRACTICES:

Name each polynomial by degree and number of terms.	Write each polynomial in standard form
1) -5	2) $12x^4 + x - 4x^3$
3) $x + 1$	4) $12 - x^3 - 3x^5 + 9x^4$
5) $8x^6 - 7$	6) $x^2 + 13x^5 + x^3 - 4x$
7) $3x^2 - x$	8) $x^5 + 2x^3 (x^2 + 2)$
9) $-8x^4 + 3x^3 - 2x^2 - 3x$	10) $(x - 5)(x + 5)$

Score: ....

Answer Key

1) Constant monomial	2) $12x^4 - 4x^3 + x$
3) Linear binomial	4) $-3x^5 + 9x^4 - x^3 + 12$
5) Sixth degree binomial	6) $13x^5 + x^3 + x^2 - 4x$
7) Quadratic binomial	8) $2x^6 + x^5 + 4x^3$
9) Quartic polynomial with four terms	10) $x^2 - 25$

Name: ....

Adding and Subtracting Polynomials

- ✓ To add polynomials, we combine like terms with some order of operations considerations thrown in.
- ✓ You have to be careful with a negative sign and don't confuse it with addition and subtraction.

EXAMPLE:

Add expressions. $(2x^3 - 6) + (9x^3 - 4x^2) = ?$

Remove parentheses: $(2x^3 - 6) + (9x^3 - 4x^2) = 2x^3 - 6 + 9x^3 - 4x^2$

Now combine like terms: $2x^3 - 6 + 9x^3 - 4x^2 = 11x^3 - 4x^2 - 6$

PRACTICES:

Simplify each expression.

1) $(x^3 + 6) - (6 + 3x^3)$	2) $(x^2 + 8) + (7x^2 - 8)$
3) $(2x^2 + x^3) - (5x^2 + 1)$	4) $(6x^2 - 4x) + (3x - 6x^2 + 1)$
5) $(x - 2x^3) - (4x^3 + 4)$	6) $(2x^3 + 2x^2) - (2x^2 - x^3 + 2)$
7) $(4x^2 - 3) + (x^2 - x^3)$	8) $(x^3 + 13x^4) - (13x^4 + 3x^3)$
9) $(x^4 + 2x^5 + 3x^3) + (4x^3 + 6x^4)$	10) $(3x^3 - 6x^6) + (3x^3 + 2x^6)$

Score: ....

Answer Key	
1) $-2x^3$	2) $8x^2$
3) $x^3 - 3x^2 + 1$	4) $-x + 1$
5) $-6x^3 + x - 4$	6) $3x^3 - 2$
7) $-x^3 + 5x^2 - 3$	8) $-2x^3$
9) $2x^5 + 7x^4 + 7x^3$	10) $-4x^6 + 6x^3$

Name: ....

Multiply and Divide Monomials

✓ When dividing monomials, we divide coefficients first and then divide their variables.

✓ If we have exponents with the same base, we'll subtract their powers.

✓ Exponent's rules:

$$x^a \times x^b = x^{a+b} , \qquad \frac{x^a}{x^b} = x^{a-b}$$

$$\frac{1}{x^b} = x^{-b}, \quad (x^a)^b = x^{a \times b}$$

$$(xy)^a = x^a \times y^a$$

EXAMPLE:

Multiply expressions. $(-3x^7)(4x^3) =?$

Use this formula: $x^a \times x^b = x^{a+b} \rightarrow x^7 \times x^3 = x^{10}$; Then: $(-3x^7)(4x^3) = -12x^{10}$

Dividing expressions. $\frac{18x^2y^5}{2xy^4} =?$

Use this formula: $\frac{x^a}{x^b} = x^{a-b}$, $\frac{x^2}{x} = x^{2-1} = x$ and $\frac{y^5}{y^4} = y^{5-4} = y$; Then: $\frac{18x^2y^5}{2xy^4} = 9xy$

PRACTICES:

Simplify.

1) $(x^3y^2)(42y^4)$	2) $\frac{100x^5y^6}{25x^6y^{11}}$
3) $(8x^4)(12x^5)$	4) $\frac{75x^{16}y^{10}}{5x^6y^7}$
5) $(-2x^{-3}y^2)^2$	6) $\frac{15x^{12}y^5}{5x^9y^2}$
7) $(11x^2y^4)(4x^9y^{10})$	8) $\frac{50x^4y^7}{25x^3y^7}$
9) $(2x^{-3}y^4)^2$	10) $\frac{-21x^8y^{13}}{3x^6y^6}$

Score: ....

Answer Key	
1) $42x^3y^6$	2) $4x^{-1}y^{-5}$
3) $96x^9$	4) $15x^{10}y^3$
5) $4x^{-6}y^4$	6) $3x^3y^3$
7) $44x^{11}y^{14}$	8) $2x$
9) $4x^{-6}y^8$	10) $-7x^2y^7$

Name: ....

Multiplying Monomials

✓ A polynomial having only one term is called monomial, like $2x$ or $7y$

EXAMPLE:

Multiply expressions. $5a^4b^3 \times 2a^3b^2 = ?$

Use this formula: $x^a \times x^b = x^{a+b}$

$a^4 \times a^3 = a^{4+3} = a^7$ and $b^3 \times b^2 = b^{3+2} = b^5$

Then: $5a^4b^3 \times 2a^3b^2 = 10a^7b^5$

PRACTICES:

Simplify each expression.

1) $2xy^2 \times 3z^2$	2) $3xyz \times 5x^2y$
3) $4pq^3 \times (-3p^3q)$	4) $s^3t^2 \times 2st^5$
5) $5p^3 \times (-2p^2)$	6) $-2p^2r \times 6pr^3$
7) $(-a)(-4a^6b)$	8) $2u^2v^3 \times (-8u^3v^3)$
9) $6u^3 \times (2u)$	10) $-5y^2 \times 4x^2y$

Score: ....

Answer Key

1) $6xy^2z^2$	2) $15x^3y^2z$
3) $-12p^4q^4$	4) $2s^4t^{10}$
5) $-10p^5$	6) $-12p^3r^4$
7) $4a^7b$	8) $-16u^5v^6$
9) $12u^4$	10) $-20x^2y^3$

Name: ....

Multiply a Polynomial and a Monomial

✓ Use the product rule for exponents to multiply monomials.

✓ Use distributive property to multiply a monomial by a polynomial.

$$a \times (b + c) = a \times b + a \times c$$

EXAMPLE:

Multiply expressions. $-4x(5x + 9) =?$

Use Distributive Property: $-4x(5x + 9) = -20x^2 - 36x$

PRACTICES:

Find each product.

1) $3(2x - 2y)$	2) $5x(4x - y)$
3) $-2x(x + 5)$	4) $11(3x + 7)$
5) $10x(5x - 2y)$	6) $4(3x - 5y)$
7) $2x(3x^3 - 5x + 4)$	8) $-4x(2 + 4xy)$
9) $3(2x^2 - 8x + 3)$	10) $-3x^2(3x^2 + 5)$

Score: ....

Answer Key	
1) $6x - 6y$	2) $20x^2 - 5xy$
3) $-2x^2 - 10$	4) $33x + 77$
5) $50x^2 - 20xy$	6) $50x^2 - 20xy$
7) $6x^4 - 10x^2 + 8x$	8) $-16x^2y - 8x$
9) $6x^2 - 24x + 9$	10) $-9x^4 - 15x^2$

Name: ....

Multiply Binomials

✓ Use "FOIL". (First-Out-In-Last)

$$(x + a)(x + b) = x^2 + (b + a)x + ab$$

EXAMPLE:

Multiply Binomials. $(x + 5)(x - 2) =$?

Use "FOIL". (First–Out–In–Last):

$(x + 5)(x - 2) = x^2 - 2x + 5x - 10$
Then simplify: $x^2 - 2x + 5x - 10 = x^2 + 3x - 10$

PRACTICES:

Multiply.

1) $(2x - 2)(x + 3)$	2) $(4x + 2)(2x + 1)$
3) $(x + 3)(x + 4)$	4) $(x^2 + 5)(x^2 - 5)$
5) $(2x - 3)(x + 4)$	6) $(2x - 6)(x + 7)$
7) $(x - 2)(3x - 4)$	8) $(2x - 5)(x + 4)$
9) $(x + 10)(x - 10)$	10) $(x - 3)(3x + 4)$

Score: ....

Answer Key	
1) $2x^2 + 4x - 6$	2) $8x^2 + 8x + 2$
3) $x^2 + 7x + 12$	4) $x^4 - 25$
5) $2x^2 + 5x - 12$	6) $2x^2 + 8x - 42$
7) $3x^2 - 10x + 8$	8) $2x^2 + 3x - 20$
9) $x^2 - 100$	10) $3x^2 - 5x - 12$

Name: ....

Factor Trinomials

✓ FOIL":
$$(x + a)(x + b) = x^2 + (b + a)x + ab$$

✓ "Difference of Squares":
$$a^2 - b^2 = (a + b)(a - b)$$
$$a^2 + 2ab + b^2 = (a + b)(a + b)$$
$$a^2 - 2ab + b^2 = (a - b)(a - b)$$

✓ "Reverse FOIL":
$$x^2 + (b + a)x + ab = (x + a)(x + b)$$

EXAMPLE:

Factor this trinomial. $x^2 - 2x - 8 =?$

Expression is broken into groups: $(x^2 + 2x) + (-4x - 8)$

Now x is a factor from $x^2 + 2x : x(x + 2)$ and factor out -4 from $-4x - 8 : -4(x + 2)$

Then: $= x(x + 2) - 4(x + 2)$, now factor out like term: $x + 2$

Then: $(x + 2)(x - 4)$

PRACTICES:

Factor each trinomial.

1) $x^2 - 12x + 27$	2) $x^2 + 5x - 24$
3) $x^2 + 13x + 30$	4) $x^2 - 81$
5) $2x^2 + 12x - 14$	6) $x^2 + 2x - 8$
7) $2x^2 + 3x + 1$	8) $2x^2 + 2x - 4$
9) $9x^2 + 3x - 2$	10) $x^2 + 15x + 56$

Score: ....

Answer Key	
1) $(x - 3)(x - 9)$	2) $(x + 8)(x - 3)$
3) $(x + 10)(x + 3)$	4) $(x + 9)(x - 9)$
5) $(x + 7)(2x - 2)$	6) $(x - 2)(x + 4)$
7) $(2x + 1)(x + 1)$	8) $(2x - 2)(x + 2)$
9) $(3x - 1)(3x + 2)$	10) $(x + 7)(x + 8)$

Name: ....

Operations with Polynomials

✓ Use distributive property to multiply a monomial by a polynomial.

$$a \times (b + c) = a \times b + a \times c$$

EXAMPLE:

Multiply. $5(2x - 6) =$

Use the distributive property: $5(2x - 6) = 5 \times 2x - 5 \times (-6) = 10x - 30$

PRACTICES:

Find each product.

1) $x^2(3x - 2)$	2) $2x^2(5x - 3)$
3) $-x(5x - 3)$	4) $x^2(-3x + 9)$
5) $5(7x + 3)$	6) $8(3x + 8)$
7) $5(10x + 4)$	8) $-3x^5(x - 3)$
9) $5(3x^2 - x + 2)$	10) $4(x^2 - 2x + 3)$

Score: ....

Answer Key

1) $3x^3 - 2x^2$	2) $10x^3 - 6x^2$
3) $-5x^2 + 3x$	4) $-3x^3 + 9x^2$
5) $35x + 15$	6) $24x + 64$
7) $50x + 20$	8) $-3x^6 + 9x^5$
9) $15x^2 - 5x + 10$	10) $4x^2 - 8x + 12$

Name: ....

Simplifying Polynomials

✓ Find "identical" terms. (They have same variables with same power).

✓ Use "FOIL". (First–Out–In–Last) for binomials:

$$(x + a)(x + b) = x^2 + (b + a)x + ab$$

✓ Using order of operation, add or subtract "identical" terms

EXAMPLE:

Simplify this expression. $(4 + x)(x - 3) =?$

Use FOIL: $(x + 4)(x - 3) = x^2 + x - 12$

PRACTICES:

Simplify each expression.

1) $-3x^2 + x^5 + 7x^5 - 2x^2 + 6$	2) $18x^5 - 3x^5 + 7x^2 - 15x^5 + 4$
3) $x(x^3 + 9) - 6(8 + x^2)$	4) $x(x^2 + 2x^3) - x^3 + x$
5) $4 - 17x^2 + 30x^2 - 17x^2 + 26$	6) $4x^2 - 8x + 3x^3 + 15x - 20x$
7) $(x - 6)(x - x^2 + 5)$	8) $(x - 5)(x + 5)$
9) $(x^4 - x) + (4x^2 - 3x^4)$	10) $x(x^2 + x + 3)$

Score: ....

Answer Key	
1) $8x^5 - 5x^2 + 6$	2) $7x^2 + 4$
3) $x^4 - 6x^2 + 9x - 48$	4) $2x^4 + x$
5) $-4x^2 + 30$	6) $3x^3 + 4x^2 - 13x$
7) $-x^3 + 7x^2 - x - 30$	8) $x^2 - 25$
9) $-2x^4 + 4x^2 - x$	10) $x^3 + x^2 + 3x$

Chapter 8 : Functions

Topics that you'll learn in this chapter:

- ➢ Relations and Functions

- ➢ Rate of change and Slope

- ➢ x and y intercept

- ➢ Slope-intercept form

- ➢ Slope-point form

- ➢ Equation of Parallel or Perpendicular lines

- ➢ Equation of Horizontal and Vertical Lines

- ➢ Function Notation

- ➢ Adding and Subtracting Functions

- ➢ Multiplying and Dividing Functions

- ➢ Composition of Functions

- ➢ Solve a Quadratic Equation

It's fine to work on any problem, so long as it generates interesting mathematics along the way – even if you don't solve it at the end of the day." – Andrew Wiles

Name:

....

Relations and Functions

✓ **RELATION:** A relationship between two sets of elements like input and output, input can have as many as outputs.

✓ **FUNCTION:** A relationship between two sets of elements like input and output and only and exactly one output related to one input.

✓ A function is a type of relation, but a relation may not be a function.

✓ **The Vertical Line Test**

A graphical method that we can observe the cross between the graph and vertical line.

The graph is a function if and only if the intersection point is one.

EXAMPLE:

Use the vertical line test to determine if the graph is function or not?

The graph is not a function

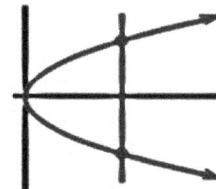

PRACTICES:

State the domain and range of each relation. Then determine whether each relation is a function.

1)

Function:

....

Domain:

....

Range:

....

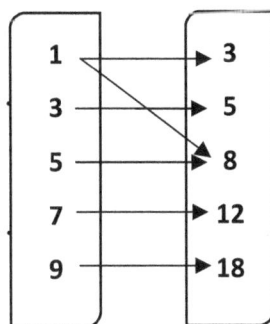

2)

Function:

....

Domain:

....

Range:

....

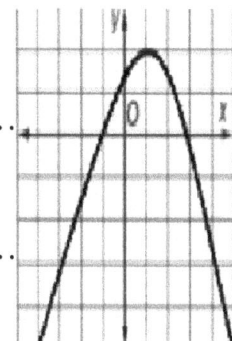

3) $\{(1, -2), (4, -1), (0, 5), (4, 0), (3, 8)\}$

Function:

....

Domain:

....

Range:

....

4)

x	y
3	4
0	1
−2	−3
6	−3
8	2

Function:

....

Domain:

....

Range:

....

Score: ....

Answer Key	
1) No, $D_f = \{1, 3, 5, 7, 9\}$, $R_f = \{3, 5, 8, 12, 18\}$	2) Yes, $D_f = (-\infty, \infty)$, $R_f = \{2, -\infty\}$
3) No, $D_f = \{1, 4, 0, 3\}$, $R_f = \{-2, -1, 5, 0, 8\}$	4) Yes, $D_f = \{3, 0, -2, 6, 8\}$, $R_f = \{4, 1, -3, 2\}$

Name: ....

Rate of change

✓ Slope can be described as "rate of change".

✓ Rate of change is a ratio between a change in one variable comparing to a corresponding change in another variable. Rate of change $= \frac{change\ in\ output\ (y)}{change\ in\ input\ (x)}$

✓ Rates of change can be positive, negative, or zero.

EXAMPLE:

The table shows the amount of money SB carwash made washing car. Find the rate of change in dollar per car?

SB Carwash	Number	4	8	12	16
	Money ($)	32	56	80	104

Rate of change $= \frac{change\ in\ output\ (y)}{change\ in\ input\ (x)} = \frac{Change\ in\ money}{Change\ in\ car} = \frac{56-32}{8-4} = \frac{24}{4} = \frac{6}{1}$, or $6 per car

PRACTICES:

What is the average rate of change of the function?

1)

Gallons	3	5	7	9
Miles	81	135	189	243

2)

Products	145	159	173	187
Costs	761	719	677	635

3)

x	4.5	6	7.5	9
y	6	15	24	33

4)

x	41	47	53	59
y	67	52	37	22

5) $f(x) = -2x + 4$, from $x = -1$ to $x = 4$?

6) $f(x) = x - 6$, from $x = -5$ to $x = 1$?

7) $f(x) = -4$, from $x = 3$ to $x = -2$?

8) $f(x) = 3x^2 + 5$, from $x = 3$ to $x = 6$?

9) $f(x) = -2x^2 - 4$, from $x = 2$ to $x = 4$?

10) $f(x) = x^3 + 3$, from $x = 1$ to $x = 2$?

Score: ....

Answer Key	
1) 27 miles per gallon	2) -3 cost per product
3) 6	4) -2.5
5) -2	6) 1
7) 0	8) 27
9) -12	10) 7

Name: ....

Slope

- ✓ The slope is used to describe the steepness and direction of lines on the coordinate plane.
- ✓ A coordinate plane is a two-dimensional plane formed by the intersection of a vertical line called y-axis and a horizontal line called x-axis. These are perpendicular lines that intersect each other at zero, and this point is called the origin.
- ✓ An ordered pair (x, y) shows the location of a point.
- ✓ A line on coordinate plane can be drawn by connecting two points.
- ✓ The slope of a line with two points A (x_1, y_1) and B (x_2, y_2) can be found by using this formula: $\frac{y_2 - y_1}{x_2 - x_1} = \frac{rise}{run}$

EXAMPLE:

Use the given points to determine the slope. Points, $(3, -7)$, $(-2, -9)$

$\text{Slope} = \frac{y_2 - y_1}{x_2 - x_1} = \frac{-9 - (-7)}{-2 + 3} = \frac{-9 + 7}{1} = \frac{-2}{1} = -2$

PRACTICES:

Find the slope of the line through each pair of points.

1) $(2, -9), (5, -6)$	2) $(14, 7), (20, 12)$
3) $(1, -5), (8, -4)$	4) $(13, -9), (15, -7)$
5) $(-5, -8), (-8, -2)$	6) $(0, 0), (12, -2)$
7) $(14, -8), (-6, 5)$	8) $(-2, 5), (-2, 8)$
9) $(-14, -9), (-6, -15)$	10) $(-19, 2), (2, -19)$

Score: ....

Answer Key	
1) 1	2) $\frac{5}{6}$
3) $\frac{1}{7}$	4) 1
5) -2	6) $-\frac{1}{6}$
7) $-\frac{13}{20}$	8) Undefined
9) $-\frac{3}{4}$	10) -1

Name: ....

x and y intercept

- ✓ x-intercept is the point at which the graph crosses the x-axis, and the value of y is zero.
- ✓ y-intercept is the point at which the graph crosses the y-axis, and the value of x is zero.

EXAMPLE:

Find the intercepts of the equation $y = 3x - 21$.

To find the x-intercept, set $y = 0$, then $0 = 3x - 21 \rightarrow 3x = 21 \rightarrow x = 7$. x-intercept $(7,0)$

To find the y-intercept, set $y = 0$, then $y = 3(0) - 21 \rightarrow y = -21$. y-intercept $(0, -12)$

PRACTICES:

Find the x and y intercepts for the following equations.

1) $5x + 3y = 15$	2) $y = x + 8$
3) $4x = y + 16$	4) $x + y = -2$
5) $4x - 3y = 7$	6) $7y - 5x + 10 = 0$
7) $\frac{3}{7}x + \frac{1}{4}y + \frac{2}{3} = 0$	8) $3x - 21 = 0$
9) $24 - 4y = 0$	10) $-2x - 6y + 42 = 12$

Score: ....

Answer Key

1) $y - intercept = 5$
$x - intercept = 3$

2) $y - intercept = 8$
$x - intercept = -8$

3) $y - intercept = -16$
$x - intercept = 4$

4) $y - intercept = -2$
$x - intercept = -2$

5) $y - intercept = -\frac{7}{3}$
$x - intercept = \frac{7}{4}$

6) $y - intercept = -\frac{10}{7}$
$x - intercept = 2$

7) $y - intercept = -\frac{8}{3}$
$x - intercept = -\frac{2}{7}$

8) $y - intercept = undefind$
$x - intercept = 7$

9) $y - intercept = 6$
$x - intercept = undefind$

10) $y - intercept = 5$
$x - intercept = 15$

Name: ....

Writing Linear Equations

✓ The equation of a line:
$$y = mx + b$$

✓ Identify the slope.

✓ Find the y-intercept. This can be done by substituting the slope and the coordinates of a point (x, y) on the line.

EXAMPLE:

If f is a linear function, with $f(4) = -2$, and $f(7) = 1$, find an equation for the function.

Slope: $\frac{y_2 - y_1}{x_2 - x_1} = \frac{1 - (-2)}{7 - 4} = \frac{1+2}{3} = 1$

$y = mx + b \rightarrow y = 1 \times x + b \rightarrow y = x + b$, and $f(7) = 1$, then $1 = 7 + b \rightarrow b = -6$

The linear equation is: $y = x - 6$

PRACTICES:

Write the equation of each line in form of $y = mx + b$.

1) $m = 2$; y-intercept$= -7$	2) $m = -\frac{2}{5}$; y-intercept$= \frac{2}{3}$
3) $f(-2) = 1; f(-3) = 4$	4) $f(6) = -1; f(2) = 7$
5) Through: $(5, 7), (3, 6)$	6) Through: $(-1.5, 2), (6.5, -2)$
7) Through: $(2, -1), (6, 11)$	8) Through: $(4, 1), (-2, 7)$
9) Through: $(2, 4), (-2, -4)$	10) Through: $(4, 5), (0, -1)$

Score: ....

Answer Key	
1) $y = 2x - 7$	2) $y = -\frac{2}{5}x + \frac{2}{3}$
3) $y = -3x - 5$	4) $y = -2x + 11$
5) $y = \frac{1}{2}x + \frac{9}{2}$	6) $y = -0.5x + 1.25$
7) $y = 3x - 7$	8) $y = -x + 5$
9) $y = 2x$	10) $y = 1.5x - 1$

Name: ....

Slope-intercept form

- ✓ The slope-intercept form is one of several ways you can write the equation of a line.
- ✓ Using the slope m and the y-intercept b, then the equation of the line is:

$$y = mx + b$$

EXAMPLE:

Solve for y when 4x - 2y = 12.

Subtract $4x$ from both sides: $-4x + 4x - 2y = 12 - 4x \rightarrow -2y = 12 - 4x$

Divide everything by -2: $y = -6 + 2x \rightarrow y = 2x - 6$

PRACTICES:

Write the slope–intercept form of the equation of each line.

1) $y - 4 = x + 3$	2) $5x + 14 = -3y$
3) $18x - 12y = -6$	4) $7x - 4y + 25 = 0$
5) $-\frac{1}{3}y = -2x + 3$	6) $5 - y - 4x = 0$
7) $-y = -6x - 9$	8) $-2(7x + y) = 24$
9) $3(y + 3) = 2(x - 3)$	10) $\frac{3}{4}y + \frac{1}{4}x + \frac{5}{4} = 0$

Score: ....

Answer Key	
1) $y = x + 7$	2) $y = -\dfrac{5}{3}x - \dfrac{14}{3}$
3) $y = \dfrac{3}{2}x + \dfrac{1}{2}$	4) $y = \dfrac{7}{4}x + \dfrac{25}{4}$
5) $y = 6x - 9$	6) $y = -4x + 5$
7) $y = 6x + 9$	8) $y = -7x - 12$
9) $y = \dfrac{2}{3}x - 5$	10) $y = -\dfrac{1}{3}x - \dfrac{5}{3}$

Name: ....

Point-slope form

✓ Using the slope m and a point (x_1, y_1) on the line, the equation of the line is:

$$(y - y_1) = m(x - x_1)$$

EXAMPLE:

Write the point-slope form of an equation of a line with a slope of 3 that passes through the point $(3, -2)$.

The slope is 3, so m = 3. We also know one point, so we know $x_1 = 3$ and $y_1 = -2$. Now we can substitute these values into the general point-slope equation.

$$y - y_1 = m(x - x_1) \rightarrow y - (-2) = \frac{3}{8}(x - 3) \rightarrow y + 2 = \frac{3}{8}(x - 3)$$

PRACTICES:

Write an equation in point–slope form for the line that passes through the given point with the slope provided.

1) $(2, -3), m = 4$	2) $(-7, 4), m = \frac{1}{5}$
3) $(0, -6), m = -2$	4) $(-a, b), m = m$
5) $(-9, 1), m = 3$	6) $(3, 0), m = -5$
7) $(-4, 11), m = \frac{1}{3}$	8) $(0, 11), (-2, 11)$
9) $\left(-\frac{1}{3}, 3\right), m = \frac{1}{5}$	10) $(0, 0), (1, -3)$

Score: ....

Answer Key	
1) $y + 3 = 4(x - 2)$	2) $y - 4 = \frac{1}{5}(x + 7)$
3) $y + 6 = -2x$	4) $y - b = m(x + a)$
5) $y - 1 = 3(x + 9)$	6) $y = -5(x - 3)$
7) $y - 11 = \frac{1}{3}(x + 4)$	8) $y - 11 = 0$
9) $y - 3 = \frac{1}{5}\left(x + \frac{1}{3}\right)$	10) $y = -3x$

Name: ....

Equation of Parallel or Perpendicular lines

- ✓ Parallel lines: The two lines will never intersect, and their slopes are identical. The only difference between the two lines is the y-intercept.
$\begin{cases} y = m_1x + b_1 \\ y = m_2x + b_2 \end{cases}$: Then, $m_1 = m_2$ and $b_1 \neq b_2$.

- ✓ Perpendicular Lines do intersect. Their intersection forms a right, or 90° angle. The slope of one line is the negative reciprocal of the slope of the other line.
$\begin{cases} y = m_1x + b_1 \\ y = m_2x + b_2 \end{cases}$: then, $m_1 = -\dfrac{1}{m_2}$, or $m_1 \times m_2 = -1$

EXAMPLE:

Given the functions below, identify the functions whose graphs are a pair of parallel lines and a pair of perpendicular lines. $\begin{cases} f(x) = 4x - 5 & g(x) = -4x + 2 \\ h(x) = \frac{1}{4}x - 5 & p(x) = 4x + 7 \end{cases}$

Parallel lines have the same slope ($f(x)$, and $p(x)$)

Perpendicular lines have negative reciprocal slopes ($g(x)$, and $h(x)$).

PRACTICES:

Write an equation of the line that passes through the given point and is parallel to the given line.	Write an equation of the line that passes through the given point and is perpendicular to the given line.
1) $(0,7), -5x - y = -4$	6) $(\frac{3}{5}, \frac{2}{5}), y = -6x - 24$
2) $(-2,-1), y = \frac{4}{5}x + 3$	7) $(-10,0), y = \frac{5}{3}x - 15$
3) $(-2,5), -8x + 5y = -18$	8) $(3,-5), y = x + 12$
4) $(3,-2), y = -\frac{2}{5}x - 3$	9) $(-3,-1), y = \frac{7}{3}x - 4$
5) $(-5,-5), 6x + 15y = -30$	10) $(0,0), y - 8x + 6 = 0$

Score: ....

Answer Key	
1) $y = -5x + 7$	2) $y = \frac{4}{5}x + \frac{3}{5}$
3) $y = \frac{8}{5}x + \frac{41}{5}$	4) $y = -\frac{2}{5}x - \frac{4}{5}$
5) $y = -\frac{2}{5}x - 7$	6) $y = \frac{1}{6}x + \frac{3}{10}$
7) $y = -\frac{3}{5}x - 6$	8) $y = -x - 2$
9) $y = -\frac{3}{7}x - \frac{16}{7}$	10) $y = -\frac{1}{8}x$

Name: ....

Equation of Horizontal and Vertical Lines

- ✓ Equations of horizontal and vertical lines only have one variable.
- ✓ The slope of horizontal lines is 0 and y-values for each point are the same. Then, the equation of horizontal lines is: $y = b$.
- ✓ The slope of vertical lines is undefined and the equation for a vertical line is: $x = a$

EXAMPLE:

Write an equation for the vertical line that passes through $(4, -1)$.

As the line is vertical, x is constant, and x always takes the same value. Then x always takes the value 4. Thus, the equation is $x = 4$.

PRACTICES:

Sketch the graph of each line.

1) $y = 3$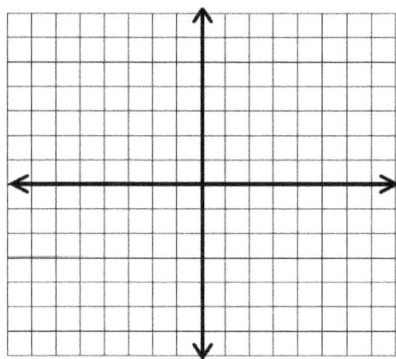

2) $y = -1$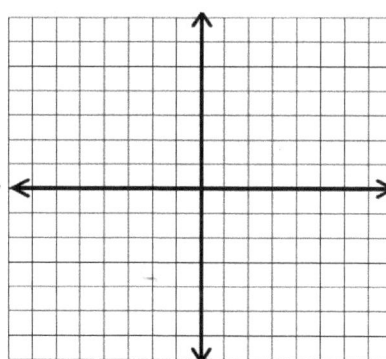

3) $x = 0$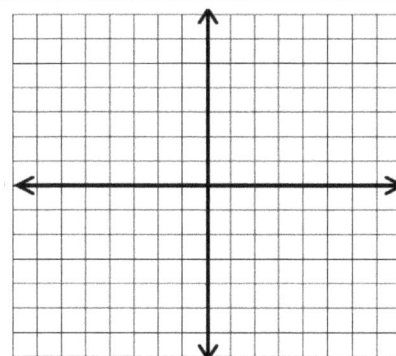

4) $x = 3$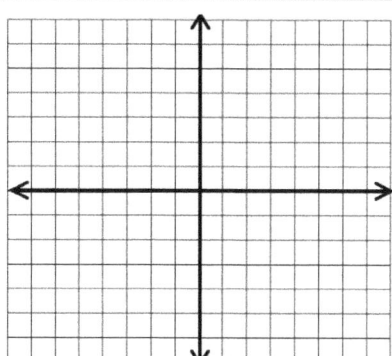

Score: ....

Answer Key

1) $y = 3$

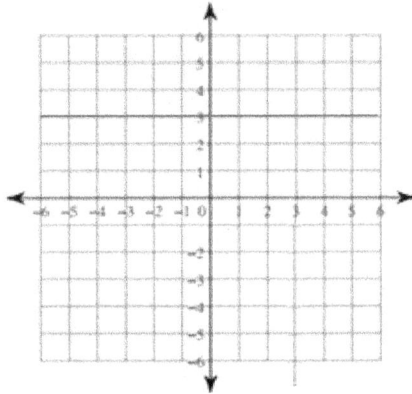

2) $y = -1$

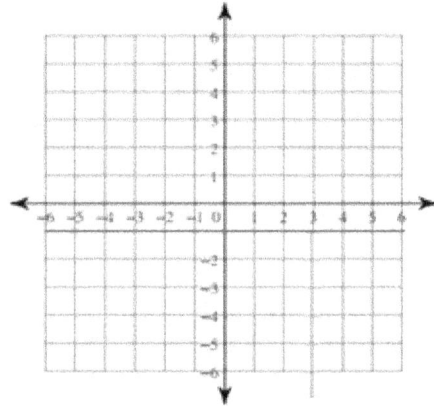

3) $x = 0$

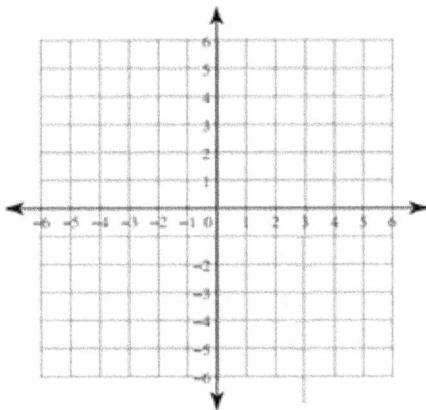

4) $x = 3$

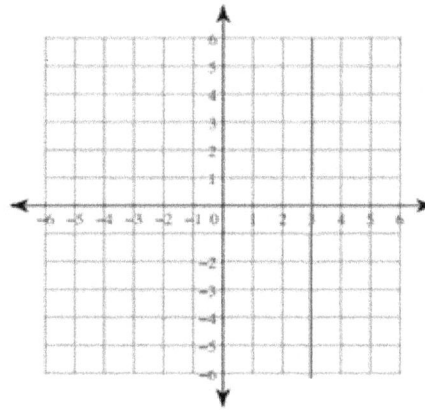

Name: ....

Function Notation

- ✓ **Function notation** is a way to write functions that is easy to read and understand. Also, it determines that a relationship is a function.
- ✓ The notation $y = f(x)$ defines a function named f. This is read y is a function of x. The letter x is the input value or independent variable. The letter y, or f (x), is the output value or dependent variable.

EXAMPLE:

Write $y = x^2 + 3x - 5$ using function notation and evaluate the function at $x = 2$.

By applying function notation, we get: $f(x) = x^2 + 3x - 5$

Evaluation: Substitute x with 2:

$f(2) = 2^2 + 3 \times 2 - 5 = 4 + 6 - 5 = 5$

PRACTICES:

Write in function notation.

1) $v = 8t$	2) $r = 4p^2 + 2p - 2$
3) $h = 15g + 8$	4) $y = 5x - \dfrac{3}{4}$

Evaluate each function.

5) $h(x) = x^3 - 8$, find $h(-2)$	6) $f(u) = 9u - 2$, find $f(^1/_3)$
7) $h(x) = 3x - 6$, find $h(a)$	8) $h(a) = -2a + 4$, find $h(3b)$
9) $h(x) = x^2 + 4x - 7$, find $h(x^2)$	10) $h(x) = x^2 + 5$, find $h(-\dfrac{a}{3})$

Score: ....

Answer Key	
1) $v(t) = 8t$	2) $r(p) = 4p^2 + 2p - 2$
3) $h(g) = 15g + 8$	4) $f(x) = 5x - \dfrac{3}{4}$
5) -16	6) 1
7) $3a - 6$	8) $-6b + 4$
9) $x^4 + 4x^2 - 7$	10) $\dfrac{1}{9}a^2 + 5$

Name: ....

Adding and Subtracting Functions

✓ Like numbers and polynomials, we can add and subtract functions which results into a new function.

✓ Let f(x) and g(x) be two functions:

We can add two functions as: $(f + g)(x) = f(x) + g(x)$

We can subtract two functions as: $(f - g)(x) = f(x) - g(x)$

EXAMPLE:

Find the sum, and difference of $f = 3x + 1$ and $g = x - 2$ at the point 5.

$f + g = (3x + 1) + (x - 2) = 4x - 1 \rightarrow (f + g)(5) = 19$

$f - g = (3x + 1) - (x - 2) = 2x + 3 \rightarrow (f - g)(5) = 13$

PRACTICES:

Perform the indicated operation.

1) $h(t) = 5t - 3$ $g(t) = 5t + 3$ Find $(h - g)(t)$.	2) $h(n) = 4n - 4$ $g(n) = n^2 - 6n + 9$ Find $(h + g)(a)$.
3) $g(a) = -3a^2 + 4$ $f(a) = 2a^2 - a + 4$ Find $(g - f)(a)$.	4) $g(x) = -x^2 + 8 - 3x$ $f(x) = 8 + 2x$ Find $(g - f)(x)$.
5) $h(x) = 3x^2 - 5$ $g(x) = -4x^2 + 2x$ Find $(h + g)(t)$.	6) $g(t) = t + 8$ $f(t) = -3t^2 + t$ Find $(g - f)(u - 1)$.
7) $h(x) = -3x + 4$ $g(x) = 2x - 6$ Find $(h + g)(2)$.	8) $k(x) = -3x + 6$ $h(x) = x^2 + 2x + 4$ Find $(k + h)(t - 3)$.

Score: ....

Answer Key	
1) -6	2) $a^2 - 2a + 5$
3) $-5a^2 + a$	4) $-x^2 - 5x$
5) $-t^2 + 2t - 5$	6) $3u^2 - 6u + 11$
7) -4	8) $t^2 - 7t + 22$

Name: ....

Multiplying and Dividing Functions

✓ Like numbers and polynomials, we can multiply and divide functions which results into a new function.

✓ Let $f(x)$ and $g(x)$ be two functions:

We can multiply two functions as: $(f \cdot g)(x) = f(x) \cdot g(x)$

We can divide two functions as: $\left(\frac{f}{g}\right)(x) = \frac{f(x)}{g(x)}$

EXAMPLE:

Given that $f(x) = x + 3$ and $g(x) = x^2 - 9$, find $(fg)(x)$ and $\left(\frac{f}{g}\right)(x)$ at the point 1.

$(fg)(x) = f(x) \times g(x) = (x+3)(x^2-9) = x^3 + 3x^2 - 9x - 27 \rightarrow (fg)(1) = -32$

$\left(\frac{f}{g}\right)(x) = \frac{f(x)}{g(x)} = \frac{x+3}{x^2-9} = \frac{x+3}{(x-3)(x+3)} = \frac{1}{x-3} \rightarrow \left(\frac{f}{g}\right)(1) = -\frac{1}{2}$

PRACTICES:

Perform the indicated operation.

1) $f(x) = x^2 - 3x$ $g(x) = 4x^2 - 2$ Find $(f \cdot g)(x)$	2) $g(t) = \frac{1}{3}t^2 + \frac{1}{3}$ $h(t) = 3t - 3$ Find $(h \cdot g)(\frac{1}{3})$
3) $f(a) = 12a - 8$ $g(a) = 5a + 10$ Find $\left(\frac{f}{g}\right)(-1)$	4) $h(a) = -2a$ $g(a) = -6a^2 - 2a$ Find $\left(\frac{h}{g}\right)(a)$
5) $g(a) = 4a - 3$ $h(a) = 4a - 2$ Find $(g \cdot h)(2)$	6) $k(n) = 2n^2 - n$ $h(n) = 3n^2 - 3$ Find $(k \cdot h)(1)$
7) $f(x) = 2x + 1$ $g(x) = 4x^2 - 1$ Find $\left(\frac{g}{f}\right)(x)$	8) $f(t) = -a + 3$ $g(t) = a^3 + 2$ Find $\left(\frac{3f}{g}\right)(a)$

Score: ....

Answer Key

1) $4x^4 - 12x^3 - 2x^2 + 6x$	2) $-\dfrac{20}{27}$
3) -4	4) $\dfrac{1}{3a+1}$
5) 30	6) 0
7) $2x - 1$	8) $\dfrac{-3a+9}{a^3+2}$

Name: ....

Composition of Functions

✓ A composite function is generally a function that is written inside another function. Composition of a function is done by substituting one function into another function.

✓ The notation used for composition is:

$$(fog)(x) = f(g(x))$$

EXAMPLE:

Using $f(x) = x + 1$ and $g(x) = 2x$, find: $(f \; o \; g)(1)$

$(f \; o \; g)(x) = f(g(x)) = 2(x + 1) = 2x + 2$
$(f \; o \; g)(1) = 4$

PRACTICES:

Using $f(x) = 2x - 8$, and $g(x) = -2x + 1$, find:

1) $f(g(0))$

2) $f(f(1))$

3) $g(f(3))$

Using $f(x) = 3x - 4a$, and $g(x) = x^2 - 2$, find:

4) $(fog)(1) = f(g(1))$

5) $(fof)(3)$

6) $(gof)(2)$

Using $f(x) = -2x + 3$, and $g(x) = x - b$, find:

7) $(fog)(-2x)$

8) $(fog)(x + 1)$

9) $(gof)(x^2)$

Score: ....

Answer Key

1) -6	2) -20
3) 5	4) $-3 - 4a$
5) $27 - 16a$	6) $16a^2 - 48a + 34$
7) $4x + 3 + 2b$	8) $-2x + 1 + 2b$
9) $-2x^2 + 3 - b$	

Name: ....

Solve a Quadratic Equation

✓ Write the equation in the Standard form:
$ax^2 + bx + c = 0$ (One side must only contain zero)
✓ Factorize the quadratic.
✓ Use quadratic formula if you couldn't factorize the quadratic.
✓ Quadratic formula: $x = \frac{-b \pm \sqrt{b^2 - 4ac}}{2a}$

EXAMPLE:

Solve: $x^2 - 4x - 21 = 0$.

$$\begin{cases} a = 1 \\ b = -4 \\ c = -21 \end{cases} \Rightarrow x = \frac{-b \pm \sqrt{b^2 - 4ac}}{2a} = \frac{-(-4) \pm \sqrt{(-4)^2 - 4(1)(-21)}}{2(1)} = \begin{cases} \frac{4 + \sqrt{100}}{2} = 7 \\ \frac{4 - \sqrt{100}}{2} = -3 \end{cases}$$

This equation is also factorable:

$$x^2 - 4x - 21 = 0 \rightarrow (x - 7)(x + 3) = 0 \rightarrow \begin{cases} x - 7 = 0 \rightarrow x = 7 \\ x + 3 = 0 \rightarrow x = -3 \end{cases}$$

PRACTICES:

Solve each equation by using the quadratic formula.

1) $x^2 + 6x = -8$	2) $3x^2 - 9x - 9 = 3$
3) $6x^2 = 24x - 18$	4) $x^2 = 3x$
5) $3x^2 + 45 = -24x$	6) $2x^2 - 24x = -72$
7) $-6x^2 - 10x - 4 = 8 - 4x^2$	8) $x^2 - 20x = -84$
9) $2x^2 + 18x + 52 = 12$	10) $x^2 + 2x = 15 + 4x$

Score: ....

Answer Key	
1) $\{-4, -2\}$	2) $\{4, -1\}$
3) $\{1, 3\}$	4) $\{3, 0\}$
5) $\{-3, -5\}$	6) $\{6\}$
7) $\{-3, -2\}$	8) $\{14, 6\}$
9) $\{-4, -5\}$	10) $\{5, -3\}$

Chapter 9 : Geometry

Topics that you'll learn in this chapter:

- ➢ The Pythagorean Theorem

- ➢ Area of Triangles and Trapezoids

- ➢ Area and Circumference of Circles

- ➢ Area and Perimeter of Polygons

- ➢ Area of Squares, Rectangles, and Parallelograms

- ➢ Volume of Cubes, Rectangle Prisms, and Cylinder

- ➢ Surface Area of Cubes, Rectangle Prisms, and Cylinder

"Mathematics is, as it were, a sensuous logic, and relates to philosophy as do the arts, music, and plastic art to poetry." — *K. Shegel*

Name: ....

The Pythagorean Theorem

✓ In any right triangle: $a^2 + b^2 = c^2$

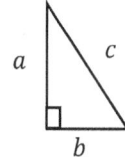

EXAMPLE:

Find the missing length.

Use Pythagorean Theorem: $a^2 + b^2 = c^2$

Then: $a^2 + b^2 = c^2 \rightarrow 3^2 + 4^2 = c^2 \rightarrow 9 + 16 = c^2$

$c^2 = 25 \rightarrow c = 5$

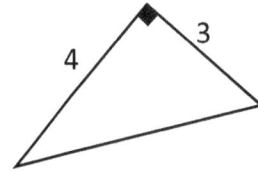

PRACTICES:

Do the following lengths form a right triangle?	Find each missing length to the nearest tenth.
1) 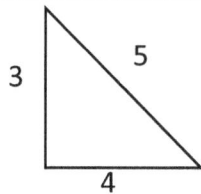	2) 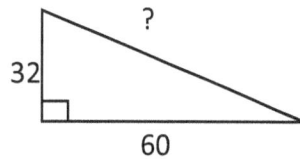
3) 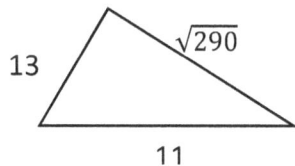	4) 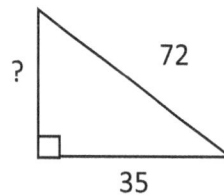
5) 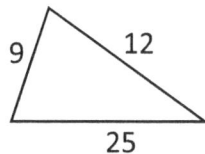	6) 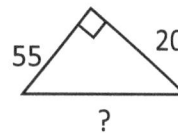
7) 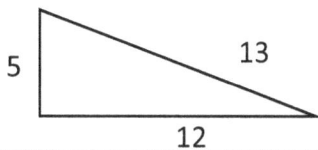	8) 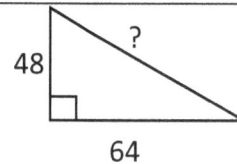

9)

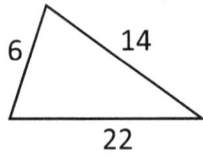

10)

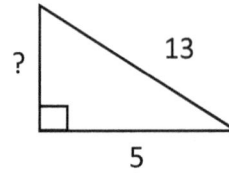

Score: ....

Answer Key

1) Yes	2) 68
3) Yes	4) 62.92
5) No	6) 58.52
7) Yes	8) 80
9) No	10) 12

Name: ....

Angles

✓ **Adjacent:** Two triangles are said to be adjacent if its two angles have common side, common vertex and do not overlap.

✓ **Vertical:** Two angles share same vertex.

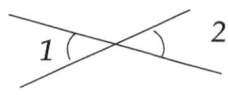

✓ **Complementary:** Sum of the measure of two complementary angles are 90°

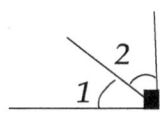

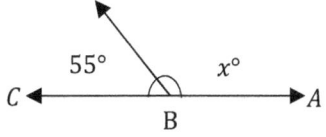

✓ **Supplementary:** Sum of the measure of two supplementary angles are 180°

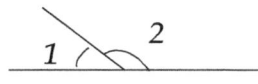

EXAMPLE:

Find x.

55° $x°$

Supplementary: Sum of the measure of two supplementary angles are 180°.

Then: $180° - 55° = 125°$

PRACTICES:

What is the value of x in the following figures?

1) 120° $x°$

2) 89° $x°$

3) 148° $x°$

4) 155° $x°$

5)

6)

7)

8)

Solve.

9) Six supplement peer to each other angles have equal measures. What is the measure of each angle? _____

10) The measure of an angle is one fourth the measure of its complementary. What is the measure of the angle? _____

Score: ....

Answer Key	
1) 60°	2) 91°
3) 32°	4) 25°
5) 50°	6) 18°
7) 20°	8) 123°
9) 30°	10) 18°

Name: ....

Area of Triangles

✓ In any triangle the sum of all angles is 180 degrees.
✓ Area of a triangle = $\frac{1}{2}(base \times height)$

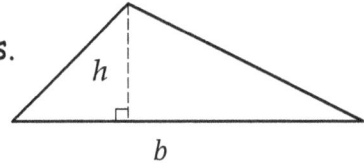

EXAMPLE:

What is the area of triangle?

Solution:

Use the are formula: Area = $\frac{1}{2}(base \times height)$

$base = 12$ and $height = 8$

Area = $\frac{1}{2}(12 \times 8) = \frac{1}{2}(96) = 48$

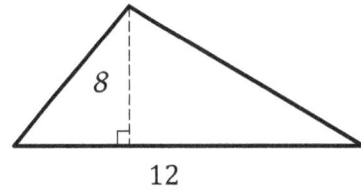

PRACTICES:

Find the area of each.

1)
c = 15 mi
h = 4 mi

2)
c = 6 m
h = 5.2 m

3)
a = 9.5 m
b = 25 m
c = 18 m
h = 9 m

4)
s = 8 m
h = 6.93 m

5)
c = 25 mi
h = 8 mi

6)
c = 10 m
h = 6.4 m

7) a = 3.5 m b = 7 m c = 16 m h = 7 m	8) s = 12 m h = 4.64 m
9) c = 13 mi h = 6 mi	10) S = 18 m h = 7.4 m

Score: ....

Answer Key

1) 30 mi²	2) 15.6 m²
3) 81 m²	4) 27.72m²
5) 100 mi²	6) 32 m²
7) 56 m²	8) 27.84 m²
9) 39 mi²	10) 133.2 m²

Name: ....

Area of Trapezoids

✓ A trapezoid is a quadrilateral with at least one pair of parallel sides.

✓ Area of a trapezoid = $\frac{1}{2}h(b_1 + b_2)$

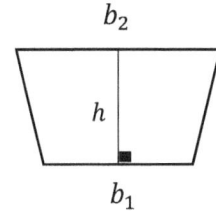

EXAMPLE:

Calculate the area of the trapezoid.

Use area formula: $A = \frac{1}{2}h(b_1 + b_2)$

$b_1 = 12$, $b_2 = 16$ and $h = 18$

Then: $A = \frac{1}{2}18(12 + 16) = 9(28) = 252\ cm^2$

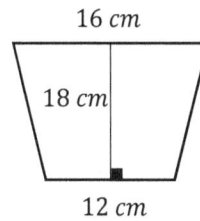

PRACTICES:

Calculate the area for each trapezoid.

1)

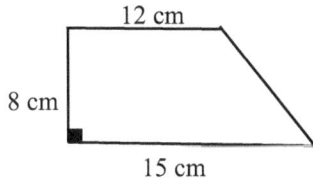

2)

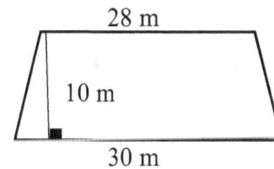

3)

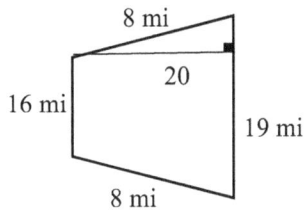

4)

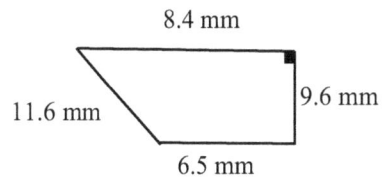

5)

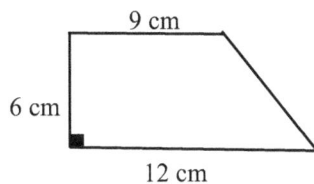

6)

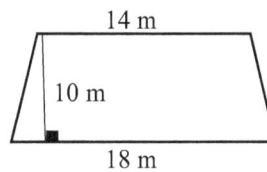

7) 7 mi 5 mi 10 mi	8) 7 mm 9.6 mm 4 mm 5 mm
9) 13 cm 10 cm 18 cm	10) 32 m 12 m 40 m

Score: ....

Answer Key

1) 108 cm^2	2) 290 m^2
3) 350 mi^2	4) 71.52 mm^2
5) 63 cm^2	6) 160 m^2
7) 42.5 mi^2	8) 24 mm^2
9) 155 cm^2	10) 432 m^2

Name: ....

Area and Perimeter of Polygons

Perimeter of a square
$= 4 \times side = 4s$

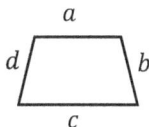

Perimeter of a rectangle
$= 2(width + length)$

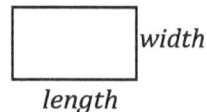

Perimeter of trapezoid
$= a + b + c + d$

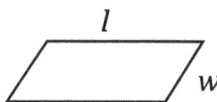

Perimeter of a regular hexagon $= 6a$

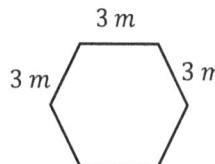

Perimeter of a parallelogram $= 2(l + w)$

EXAMPLE:

Find the perimeter of following regular hexagon.

Perimeter of Pentagon $= 6a$

Perimeter of Pentagon $= 6a = 6 \times 3 = 18m$

(regular hexagon with sides labeled 3 m, 3 m, 3 m, 3 m)

PRACTICES:

Find the area and perimeter of each.

1)
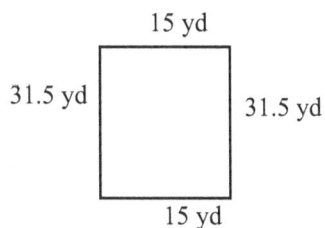
15 yd
31.5 yd   31.5 yd
15 yd

2)
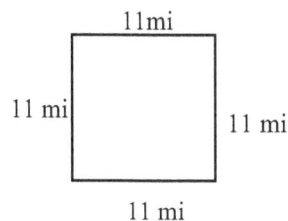
11mi
11 mi   11 mi
11 mi

3)
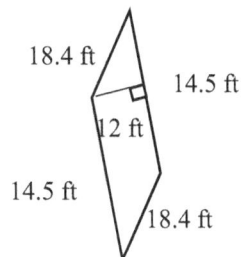
18.4 ft
14.5 ft
12 ft
14.5 ft
18.4 ft

4)
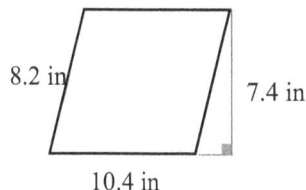
8.2 in   7.4 in
10.4 in

5) 15 cm → ← 10 cm 13 cm	6) 5 mm 8 mm 6 mm 5 mm

Find the perimeter of each shape.

7) 6 m 6 m 6 m	8) 11mm 11 mm
9) 13 ft 13 ft	10) 20 in 19 in

Score: ....

Answer Key

1) Area: 472.5 yd^2, Perimeter: 93 yd	2) Area: 121 mi^2, Perimeter: 44 mi
3) Area: 174 ft^2, Perimeter: 65.8 ft	4) Area: 76.96 in^2, Perimeter: 37.2 in
5) Area: $75cm^2$, Perimeter 52 cm	6) Area: 70 mm^2, Perimeter:38 mm
7) P: 36 m	8) P: 44 mm
9) P: 52 ft	10) P: 78 in

Name: ....

Area and Circumference of Circles

- ✓ We use variable r for the radius and d for diameter in a circle and π is about 3.14.
- ✓ Area of a circle$= \pi r^2$
- ✓ Circumference of a circle$= 2\pi r$

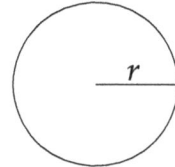

EXAMPLE:

Find the Circumference and area of the circle.

Use Circumference formula: $Circumference = 2\pi r$

$r = 6$, then: $Circumference = 2\pi(6) = 12\pi$

$\pi = 3.14$ then: $Circumference = 12 \times 3.14 = 37.68$

Use area formula: $Area = \pi r^2$,

$r = 6$ then: $Area = \pi(6)^2 = 36\pi$, $\pi = 3.14$ then: $Area = 36 \times 3.14 = 113.04$

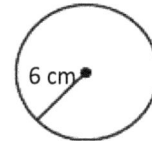

PRACTICES:

Find the area and circumference of each. $(\pi = 3.14)$

1)

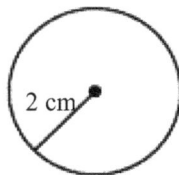

2)

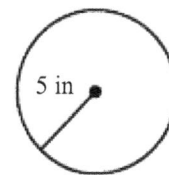

3)

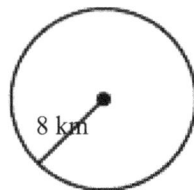

4)

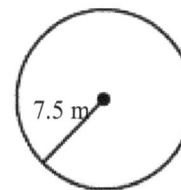

5)

6)

4 m

10 cm

7)

8)

2.5cm

1.5 in

9)

10)

6 km

24 in

Score: ....

Answer Key

1) Area: 12.56 cm², Circumference: 12.56cm.	2) Area: 78.5 in2, Circumference: 31.4 in.
3) Area: 200.96 km², Circumference: 50.24 km.	4) Area: 176.625 m2, Circumference: 47.1 m.
5) Area: 50.24 m², Circumference: 25.12 m	6) Area: 78.5 cm2, Circumference: 31.4 cm.
7) Area: 4.906 cm², Circumference: 7.85 cm.	8) Area: 1.766 in2, Circumference: 4.71 in.
9) Area: 113.04 km², Circumference:37.68 km.	10) Area: 452.16 in², Circumference: 75.36 in

Name: ....

Volume of Cubes

- ✓ A three-dimensional solid object bounded by six square sides is called cube.
- ✓ The measure of the amount of space inside of a solid figure is called volume, like a cube, ball, cylinder, or pyramid.
- ✓ Volume of a cube = $(one\ side)^3$

EXAMPLE:

Find the volume of this cube.

Use volume formula: $volume = (one\ side)^3$

Then: $volume = (one\ side)^3 = (2)^3 = 8\ cm^3$

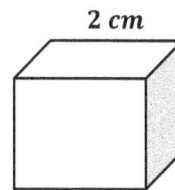

2 cm

PRACTICES:

Find the volume of each.

1)

2)

3)

4)

5)

6)

7) 4 ft	8) 6 m
9) 5 in	10) 3 miles

Score: ....

Answer Key

1) 6	2) 34
3) 7	4) 6
5) 41	6) 54
7) $64\,ft^3$	8) $216\,m^3$
9) $125\,in^3$	10) $27\,mi^3$

Name: ....

Volume of Rectangle Prisms

✓ A solid 3-dimensional object which has six rectangular faces.

✓ Volume of a Rectangular prism = $Length \times Width \times Height$

   Volume = $l \times w \times h$

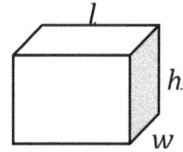

EXAMPLE:

Find the volume and surface area of rectangular prism.

   Use volume formula: $Volume = l \times w \times h$

   Then: $Volume = 10 \times 5 \times 8 = 400 \ m^3$

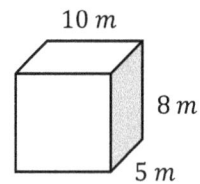

PRACTICES:

Find the volume of each of the rectangular prisms.

1) 10 cm, 12cm, 7 cm

2) 11 cm, 9 cm, 2 cm

3) 4 m, 4 m, 4 m

4) 15 cm, 19 cm, 5 cm

5) 17.5, 10 cm, 4cm

6) 5.5 m, 5.5 m, 5.5 m

7) 7 m 7 m 7 m	8) 20 ft 11 ft 3 ft
9) 12.5 km 8 km 3 km	10) 8 in 10 in 5 in

Score: ....

Answer Key	
1) 840 cm^3	2) 198 cm^3
3) 64 m^3	4) 1,425 cm^3
5) 700 cm^3	6) 166.375 cm^3
7) 343 cm^3	8) 660 ft^3
9) 300 km^3	10) 400 in^3

Name: ....

Surface Area of Cubes

✓ A three-dimensional solid object bounded by six square sides is called cube.

surface area of cube $= 6 \times (one\ side)^2$

EXAMPLE:

Find the volume and surface area of this cube.

surface area of cube: $6(one\ side)^2 = 6(2)^2 = 6(4) = 24\ cm^2$

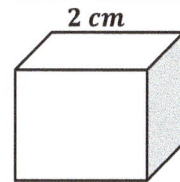

2 cm

PRACTICES:

Find the surface of each cube.

1) 　　　　　　7 mm

2) 　　　　　　10.5 mm

3) 　　　　　　3.5 cm

4) 　　　　　　4 m

5)

　　　3.2 in

6) 　　　　　　8.1 ft

7) 1.6 in	8) 11 m
9) 5.2 in	10) 2.25 mm

Score: ....

Answer Key

1) 294 mm²	2) 661.5 mm²
3) 73.5 cm²	4) 96 m²
5) 61.44 in²	6) 393.66 ft²
7) 15.36 in²	8) 726 m²
9) 162.24 in²	10) 30.375 mm²

Name: ....

Surface Area of a Rectangle Prism

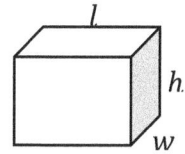

- ✓ A solid 3-dimensional object which has six rectangular faces.
  Surface area= $2(wh + lw + lh)$

EXAMPLE:

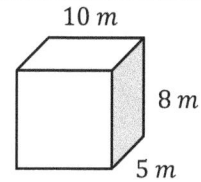

Find the volume and surface area of rectangular prism.

Use surface area formula: $Surface\ area = 2(wh + lw + lh)$

Then: $Surface\ area = 2(5 \times 8 + 10 \times 5 + 10 \times 8) =$

$2(40 + 50 + 80) = 340\ m^2$

PRACTICES:

Find the surface of each prism.

1)

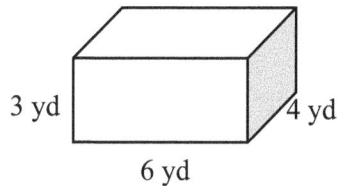

3 yd, 4 yd, 6 yd

2)

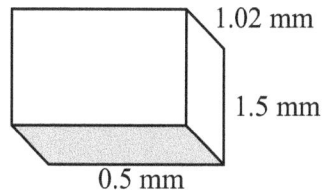

1.02 mm, 1.5 mm, 0.5 mm

3)

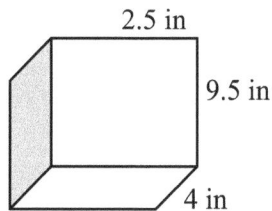

2.5 in, 9.5 in, 4 in

4)

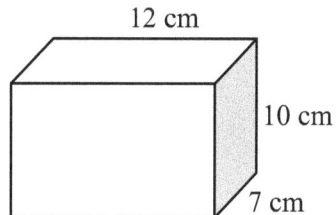

12 cm, 10 cm, 7 cm

5)

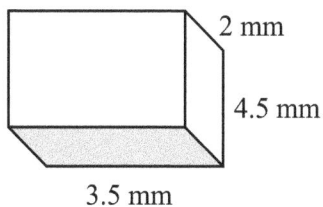

2 mm, 4.5 mm, 3.5 mm

6)

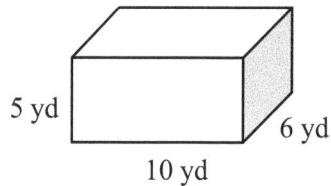

5 yd, 6 yd, 10 yd

7) 14 cm 8 cm 5 cm	8) 2 yd 6 yd 5 yd
9) 3.5 in 7.5 in 6 in	10) 4 mm 8 mm 6 mm

Score: ....

Answer Key

1) 108 yd^2	2) 5.58 mm^2
3) 143.5 in^2	4) 548 cm^2
5) 63.5 mm^2	6) 280 yd^2
7) 444 cm^2	8) 104 yd^2
9) 184.5 in^2	10) 208 mm^2

Name: ....

Volume of a Cylinder

- ✓ A solid geometric figure with straight parallel sides and a circular cross section is called a cylinder.
- ✓ Volume of Cylinder Formula $= \pi(radius)^2 \times height$
  $\pi = 3.14$

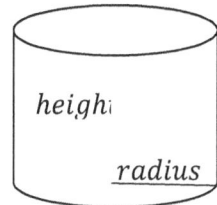

height

radius

EXAMPLE:

Find the volume of the follow Cylinder.

Use volume formula: $Volume = \pi(radius)^2 \times height$

Then: $Volume = \pi(4)^2 \times 6 = \pi 16 \times 6 = 96\pi$

$\pi = 3.14$ then: $Volume = 96\pi = 301.44$

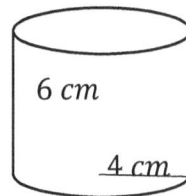

6 cm

4 cm

PRACTICES:

Find the volume of each cylinder. ($\pi = 3.14$).

1)

4 in

6 in

2)

7 m

10 m

3)

3 m

6 m

4)

2 in

4.5 in

5)

7.5 m

4 m

6)

14 in

3.5 in

7) 10 in / 7.5 in	8) 6 ft / 10 ft
9) 5 in / 9 in	10) 12 yd / 2 yd

Score: ....

Answer Key	
1) 301.44 in³	2) 1538.6 m³
3) 42.39 m³	4) 14.13 in³
5) 376.8 m³	6) 538.51 in³
7) 588.75 in³	8) 282.6 ft³
9) 706.5 in³	10) 150.72 yd³

Name: ....

Surface Area of a Cylinder

- ✓ A solid geometric figure with straight parallel sides and a circular cross section is called cylinder.
- ✓ Surface area of a cylinder= $2\pi r^2 + 2\pi rh$

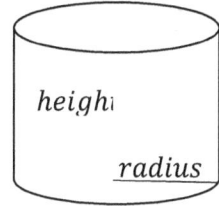

height

radius

EXAMPLE:

Find the Surface area of the follow Cylinder.

Use surface area formula: $Surface\ area = 2\pi r^2 + 2\pi rh$

Then: $= 2\pi(4)^2 + 2\pi(4)(6) = 2\pi(16) + 2\pi(24) = 32\pi + 48\pi = 80\pi$

$\pi = 3.14$ then: $Surface\ area = 80 \times 3.14 = 251.2$

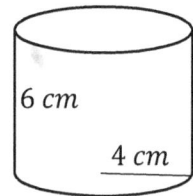

6 cm

4 cm

PRACTICES:

Find the surface of each cylinder. ($\pi = 3.14$).

1) 5 ft, 8 ft

2) 7 cm, 4 cm

3) 6 in, 10 in

4) 2 yd, 5.5 yd

5) 18 in, 12 in

6) 1.5 m, 4 m

| 7) 6 in · 10 in | 8) 8 ft · 4 ft |
| 9) 1.4 yd 6.6 yd | 10) 10 in · 20 in · |

Score: ....

Answer Key

1) 226.08 ft²	2) 113.04 cm²
3) 224.92 in²	4) 94.2 yd²
5) 1,186.92 in²	6) 51.81 m²
7) 244.92 in²	8) 502.4 ft²
9) 70.336 yd²	10) 785 in²

Chapter 10 : Statistics

Topics that you'll learn in this chapter:

- ➤ Mean, Median, Mode, and Range of the Given Data

- ➤ Box and Whisker Plots

- ➤ Bar Graph

- ➤ Stem– And– Leaf Plot

- ➤ The Pie Graph or Circle Graph

- ➤ Dot and Scatter Plots

- ➤ Probability of Simple Events

"The book of nature is written in the language of Mathematic" -Galileo

Name: ....

Mean and Median

- ✓ Mean: $\dfrac{\text{sum of the data}}{\text{total number of data entires}}$

- ✓ Median: Middle value in the sorted list of numbers

- ✓ When there are two middle numbers, we average them

EXAMPLE:

What is the median of these numbers? $4, 9, 13, 8, 15, 18, 5, 11$

Write the numbers in order: $4, 5, 8, 9, 11, 13, 15, 18$

Median is the number in the middle. Therefore, there are 9 *and* 11 in the middle, then

find the average: $\dfrac{9+11}{2} = \dfrac{20}{2} = 10$, the median is 10

PRACTICES:

Find Mean and Median of the Given Data.

1) $8, 10, 7, 3, 12$	2) $4, 6, 9, 7, 5, 19$
3) $5, 11, 1, 1, 8, 9, 20$	4) $12, 4, 2, 7, 3, 2$
5) $3, 5, 7, 4, 7, 8, 9$	6) $5, 10, 4, 4, 9, 12, 9$
7) $10, 4, 8, 5, 9, 6, 7, 19$	8) $16, 3, 4, 3, 7, 6, 18$

Solve.

9) In a javelin throw competition, five athletics score 23, 45, 53, 53, 13 and 61 meters. What are their Mean and Median? _____

10) Eva went to shop and bought 7 apples, 4 peaches, 6 bananas, 3 pineapples and 4 melons. What are the Mean and Median of her purchase? _____

Score: ....

Answer Key	
1) Mean: 8, Median: 8	2) Mean: 8.33, Median: 6.5
3) Mean: 7.85, Median: 8	4) Mean: 5, Median: 3.5
5) Mean: 6.14, Median: 7	6) Mean: 7.57, Median: 9
7) Mean: 8.5, Median: 7.5	8) Mean: 8.14, Median: 6
9) Mean: 39.106, Median: 45	10) Mean: 4.8, Median: 4

Name: ....

Mode and Range

✓ Mode: The most appeared value in the list.

✓ Range: The difference of highest value and lowest value in the list

EXAMPLE:

What is the mode(s) of these numbers? **22, 16, 12, 9, 7, 6, 4, 6, 9**

Mode: The most appeared value in the list.

Therefore: modes are 6 and 9

PRACTICES:

Find Mode and Rage of the Given Data.

1) 10, 12, 8, 8,4, 1, 9 Mode: _____Range: _____	2) 4, 6, 4, 13, 2, 13, 19, 13 Mode: _____Range: _____
3) 8, 8, 7, 2, 7, 7, 5, 6, 5 Mode: _____Range: _____	4) 12, 9, 12,6, 12, 9, 10 Mode: _____Range: _____
5) 2, 2, 4, 3, 2, 10, 8 Mode: _____Range: _____	6) 6, 1, 4, 20, 19, 2, 7, 1, 5, 1 Mode: _____Range: _____
7) 16, 35, 9, 7, 7, 5, 14, 13, 7 Mode: _____Range: _____	8) 7, 6, 6, 9, 16, 6, 7, 5 Mode: _____Range: _____

Solve.

9) A stationery sold 15 pencils, 26 red pens, 22 blue pens, 10 notebooks, 12 erasers, 22 rulers and 42 color pencils. What are the Mode and Range for the stationery sells?

Mode: _____ Range: _____

10) In an English test, eight students score 24, 13, 17, 21, 19, 13, 13 and 17. What are their Mode and Range? _____

Score: ....

Answer Key

1) Mode: 8, Range: 11	2) Mode: 13, Range: 17
3) Mode: 7, Range: 6	4) Mode: 12, Range: 6
5) Mode: 2, Range: 8	6) Mode: 1, Range: 19
7) Mode: 7, Range: 30	8) Mode: 6, Range: 11
9) Mode: 22, Range: 32	10) Mode: 13, Range: 11

Name: ....

Times Series

✓ A precise representation of the distribution of numerical data is referred as Time Series.

EXAMPLE:

Use the following Graph to complete the table.

Day	Distance (km)
1	
2	

Answer:

Day	Distance (km)
1	359
2	460
3	278
4	547
5	360

→

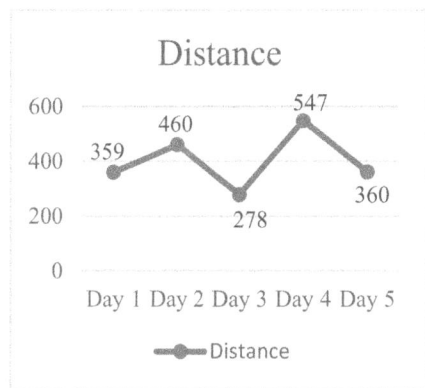

PRACTICES:

Use the following Graph to complete the table.

1)

Day	Distance (km)
1	
2	

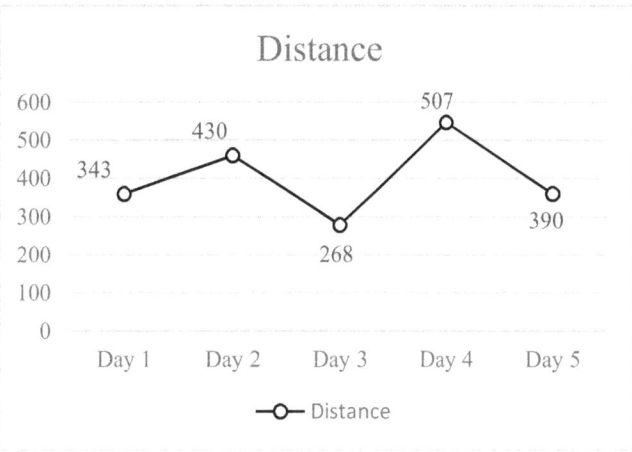

The following table shows the number of births in the US from 2007 to 2012 (in millions).

2)

Draw a time series for the table.

Year	Number of births (in millions)
2007	6.42
2008	6.45
2009	6.33
2010	5.9
2011	4.35
2012	4.35

Score: ....

Answer Key

1)

Day	Distance (km)
1	343
2	430
3	268
4	507
5	390

2)

Number of Births

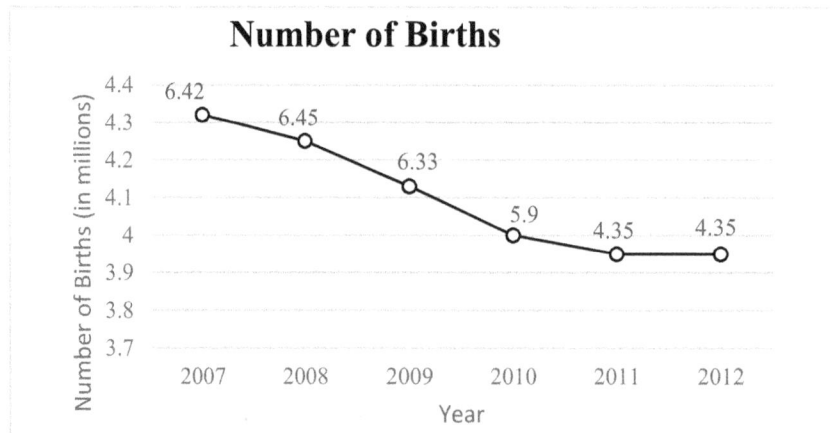

Name: ....

Box and Whisker Plot

✓ Box-and-whisker plots display data including quartiles.

✓ IQR – interquartile range shows the difference from Q1 to Q3.

✓ Extreme Values are the lowest and highest values in a data set.

EXAMPLE:

$73, 84, 86, 95, 68, 67, 100, 94, 77, 80, 62, 79$

Maximum: 100, Minimum: 62; Q_1: 70.5; Q_2: 79.5; Q_3: 90

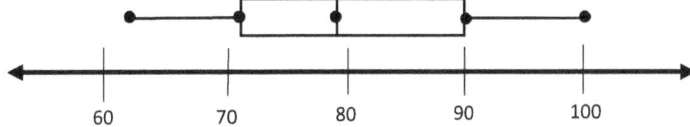

PRACTICES:

Make box and whisker plots for the given data.

1) $1, 5, 20, 8, 3, 10, 13, 11, 14, 17, 18, 15, 23$

2) $2, 7, 23, 11, 13, 9, 16, 5, 18, 22, 20, 17, 19$

3) $3, 7, 9, 10, 11, 5, 14, 19, 20, 21, 22, 8, 14$

4) $4, 6, 5, 15, 12, 14, 10, 7, 21, 17, 8, 22, 6$

Score: ....

Answer Key

1) 1,3, 5, 8, 10, 11, 13, 14, 15, 17,18, 20, 23

   Maximum: 23, Minimum: 1, Q_1: 8, Q_2: 13, Q_3: 17

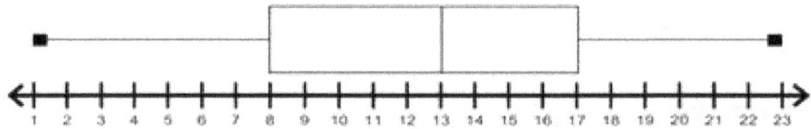

2) 2, 7, 23, 11, 13, 9, 16, 5, 18, 22, 20, 17, 19

   Maximum: 23, Minimum: 2, Q_1: 9, Q_2: 16, Q_3: 19

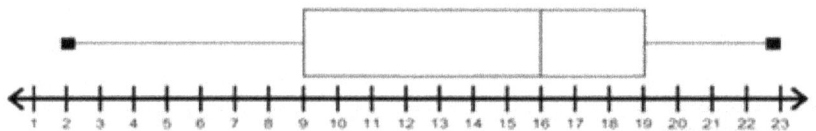

3) 3,7, 9, 10, 11, 5, 14, 19, 20, 21 ,22, 8, 14

   Maximum: 22, Minimum: 2, Q_1: 8, Q_2: 11, Q_3: 19

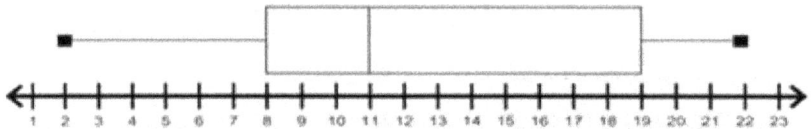

4) 4, 6, 5, 15, 12, 14, 10, 7, 21, 17, 8, 22, 6

   Maximum: 22, Minimum: 4, Q_1: 6, Q_2: 10, Q_3: 15

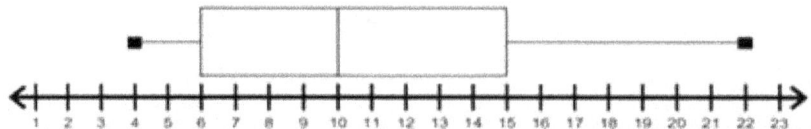

Name: ....

Bar Graph

✓ A chart that presents data with bars in different heights to match with the values of the data is called a bar graph. We can graph the bars vertically or horizontally.

EXAMPLE:

Graph the given information as a bar graph.

Name of the Sport	Total Number of Students
Football	15
Volleyball	7
Table Tennis	7
Basketball	12
Chess	9

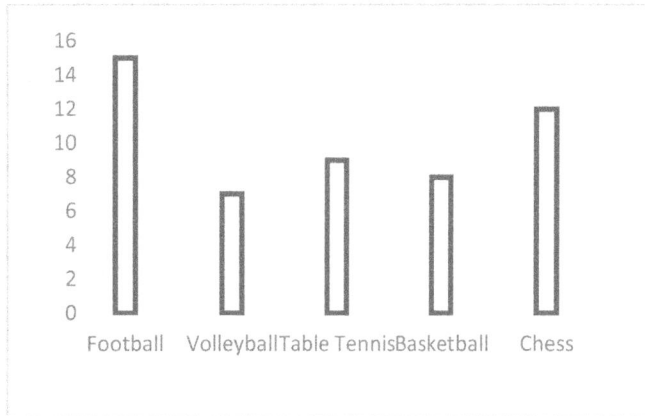

PRACTICES:

Graph the given information as a bar graph.

1)

Day	Sale House
Monday	6
Tuesday	4
Wednesday	10
Thursday	5
Friday	2
Saturday	8
Sunday	1

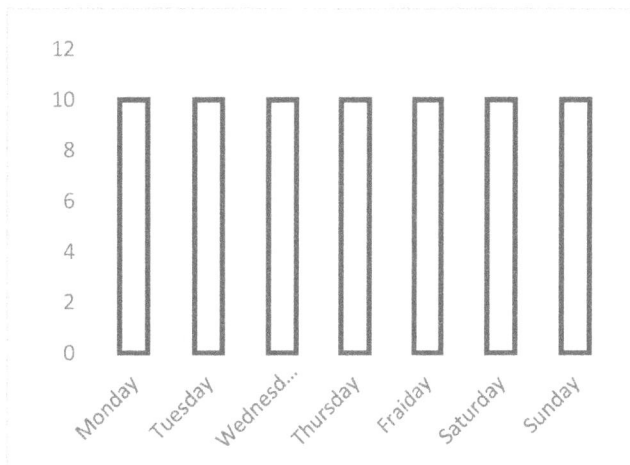

2)

Day	Sale House
Monday	8
Tuesday	6
Thursday	3
Friday	10
Saturday	4
Sunday	2

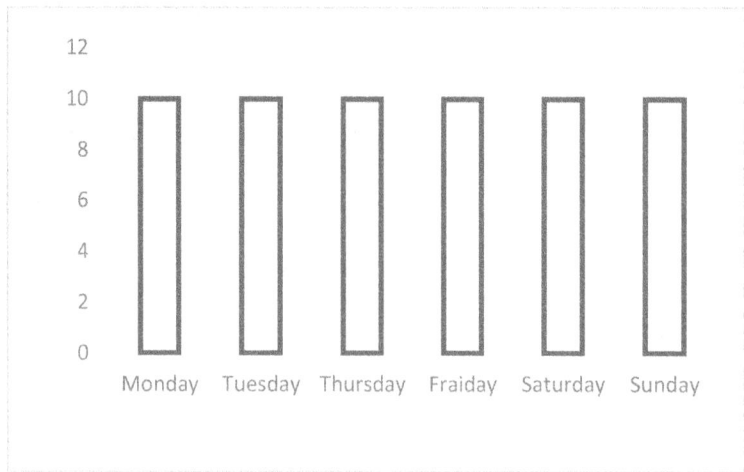

Score: ....

Answer Key

1)

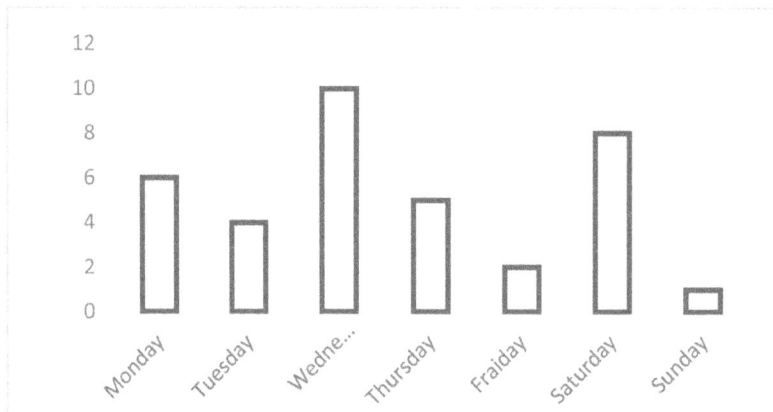

2)

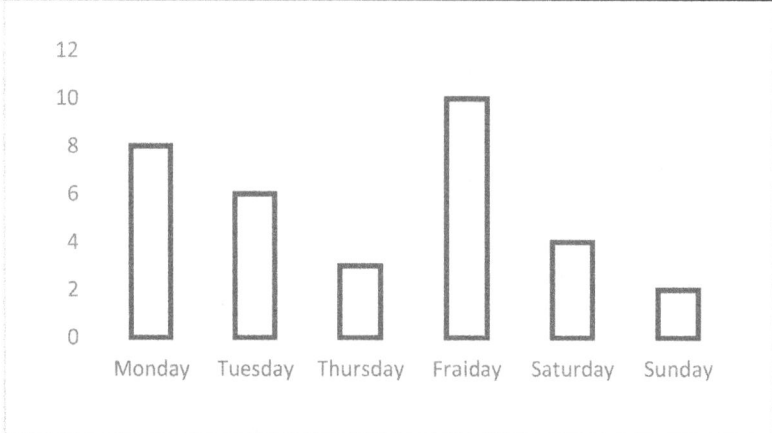

Dot plots

✓ The representation of a distribution that consists of group of data points plotted on a simple scale is referred as a dot plot. Dot plots are used for continuous, univariate and quantitative data. If there are few data points, then they can be labelled.

✓ Dot plots are one of the simplest statistical plots and are suitable for small to moderate sized data sets.

EXAMPLE:

A survey of "How many books each student purchased?" has these results How many students purchase 4 books?

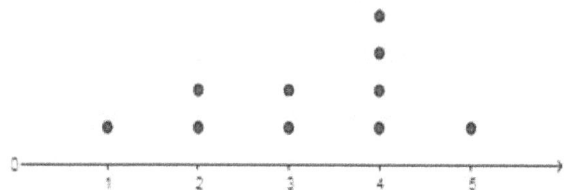

4 students

PRACTICES:

A survey of "How many pets each person owned?" has these results:

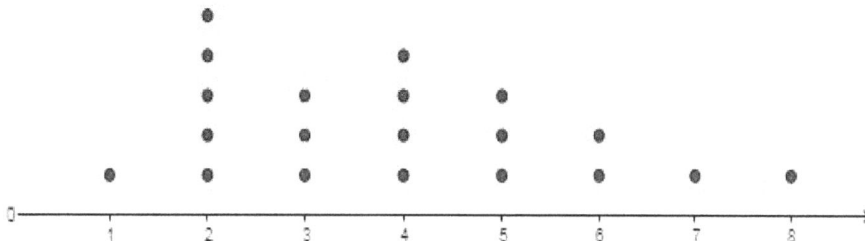

1) How many people have at least 3 pets?

2) How many people have 2 and 3 pets?

3) How many people have 4 pets?

4) How many people have 2 or less than 2 pets?

5) How many people have more than 7 pets?

6) How many people have more than 4 pets?

Score: ....

Answer Key

1) 4	2) 8
3) 4	4) 6
5) 1	6) 7

Name: ....

Scatter Plots

✓ The values with points that represent the relationship between two sets of data are shown by a scatter(x, y) plot.

✓ The horizontal values are taken as x and vertical data is taken as y.

EXAMPLE:

Construct a scatter plot.

x	1	2	3	4	5
y	2	4.5	1.5	5	2

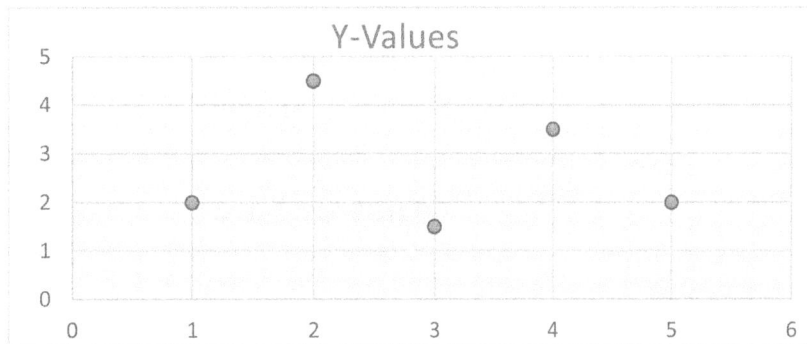

PRACTICES:

Construct a scatter plot.

1)

x	1	2.5	3	3.5	4	5
y	4	3.5	4.5	2.5	8	2

2)

x	1	2	3.5	4	4.5	5
y	3	1	2.5	1.5	1.5	1

Score: ....

Answer Key

1)

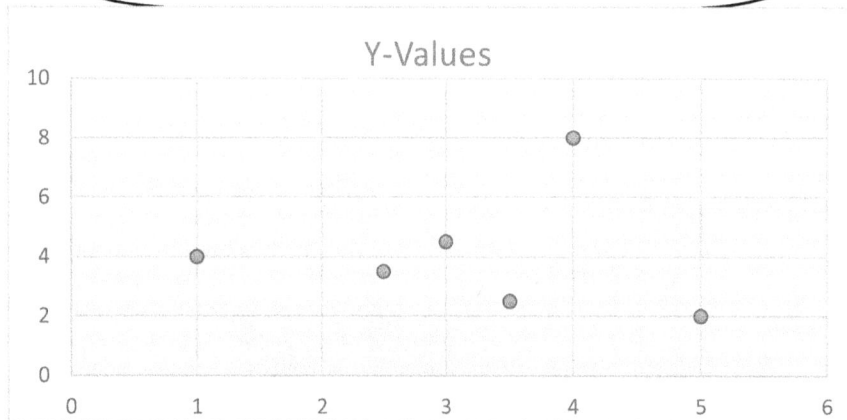

2)

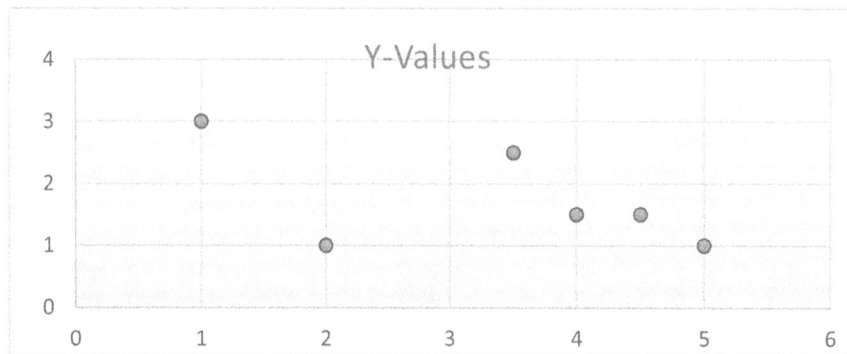

Stem–And–Leaf Plot

✓ Stem-and-leaf plots display the frequency of the values in a data set.

✓ We can make a frequency distribution table for the values, or we can use a stem-and-leaf plot.

EXAMPLE:

$56, 58, 42, 48, 66, 64, 53, 69, 45, 72$

Stem	leaf
4	2 5 8
5	3 6 8
6	4 6 9
7	2

PRACTICES:

Make stem ad leaf plots for the given data.

1) $42, 14, 17, 21, 44, 24, 18, 47, 23, 24, 19, 12$

Stem | Leaf plot

2) $10, 65, 14, 18, 69, 11, 33, 61, 66, 38, 15, 35$

Stem | Leaf plot

3) $122, 87, 99, 86, 100, 126, 92, 129, 88, 121, 91, 107$

Stem | Leaf plot

4) $60, 51, 119, 69, 72, 59, 110, 65, 77, 59, 65, 112, 71$

Stem | Leaf plot

Score: ....

Answer Key

1)

Stem	leaf
1	2 4 7 8 9
2	1 3 4
4	2 4 7

2)

Stem	leaf
1	0 1 4 5 8
3	3 5 8
6	1 5 6 9

3)

Stem	leaf
8	6 7 8
9	1 2 9
10	0 7
12	1 2 6 9

4)

Stem	leaf
5	1 9 9
6	0 5 5 9
7	1 2 7
11	0 2 9

Name: ....

The Pie Graph or Circle Graph

✓ A circular chart divided into sectors is said to be a pie chart; each sector represents the relative size of each value.

EXAMPLE:

A library has 840 books that include Mathematics, Physics, Chemistry, English, and History. Use following graph to answer question.

What is the number of Mathematics books?

Number of total books = 840,

Percent of Mathematics books = 30% = 0.30

Then: $0.30 \times 840 = 252$

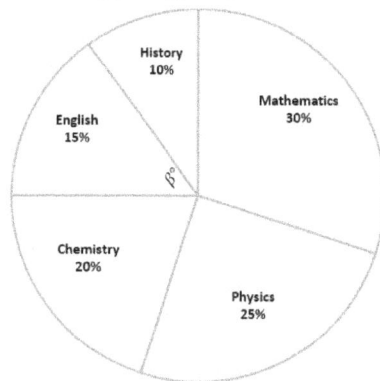

History 10%
Mathematics 30%
English 15%
Chemistry 20%
Physics 25%

PRACTICES:

Favorite Sports:

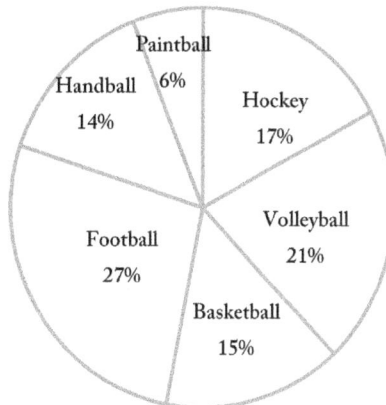

Paintball 6%
Handball 14%
Hockey 17%
Football 27%
Volleyball 21%
Basketball 15%

SPORTS

1) What percentage of pie graph is paintball?

2) What percentage of pie graph is Hockey and volleyball?

3) What percentage of pie not Football and Basketball?

4) What percentage of pie graph is Hockey and Handball and Football?

5) What percentage of pie graph is Basketball?

6) What percentage of pie not Handball and Paintball?

Score: ....

Answer Key

1) 6%	2) 38%
3) 58%	4) 58%
5) 15%	6) 80%

Name: ....

Probability of Simple Events

✓ Probability is the possibility of something happening in the future. It is shown as a number between zero (can never happen) to 1 (will always happen).

✓ Probability can be written as a fraction, a decimal, or a percent.

EXAMPLE:

If there are 8 red balls and 12 blue balls in a basket, what is the probability that John will pick out a red ball from the basket?

There are 8 red ball and 20 are total number of balls. Therefore, probability that John will pick out a red ball from the basket is 8 out of 20 or $\frac{8}{8+12} = \frac{8}{20} = \frac{2}{5}$

PRACTICES:

Solve.

1) A number is chosen at random from 28 to 35. Find the probability of selecting factors of 5.

2) A number is chosen at random from 1 to 60. Find the probability of selecting multiples of 15.

3) Find the probability of selecting 4queens from a deck of card.

4) A number is chosen at random from 8 to 19. Find the probability of selecting factors of 3.

5) What probability of selecting a ball less than 6 from 10 different bingo balls?

6) A number is chosen at random from 1 to 10. What is the probability of selecting a multiple of 2?

7) A card is chosen from a well-shuffled deck of 52 cards. What is the probability that the card will be a king OR a queen?

8) A number is chosen at random from 1 to 20. What is the probability of selecting multiples of 5?

9) A number is chosen at random from 1 to 10. Find the probability of selecting number 4 or smaller numbers.

10) Bag A contains 9 red marbles and 3 green marbles. Bag B contains 9 black marbles and 6 orange marbles. What is the probability of selecting a green marble at random from bag A? What is the probability of selecting a black marble at random from Bag B?

Score: ....

Answer Key

1) $\frac{1}{4}$	2) $\frac{1}{15}$
3) $\frac{1}{13}$	4) $\frac{1}{3}$
5) $\frac{1}{2}$	6) $\frac{1}{2}$
7) $\frac{2}{13}$	8) $\frac{1}{5}$
9) $\frac{2}{5}$	10) $\frac{1}{4}, \frac{3}{5}$

Chapter 11 : AFOQT Math Practice Tests

The Air Force Officer Qualifying Test (AFOQT) is a standardized test to assess skills and personality traits that have proven to be predictive of success in officer commissioning programs such as the training program.

The AFOQT is used to select applicants for officer commissioning programs, such as Officer Training School (OTS) or Air Force Reserve Officer Training Corps (Air Force ROTC) and pilot and navigator training.

The AFOQT is a multiple-aptitude battery that measures developed abilities and helps predict future academic and occupational success in the military. The AFOQT is a multiple-choice test which consists of 12 subtests and two of them are Arithmetic Reasoning and Mathematics Knowledge.

In this section, there are 2 complete Arithmetic Reasoning and Mathematics Knowledge AFOQT Tests. Take these tests to see what score you'll be able to receive on a real AFOQT test.

Good Luck

AFOQT Practice Test 1

Arithmetic Reasoning

- ○ **25 questions**

- ○ **Total time for this section: 29 Minutes**

- ○ **Calculators are not allowed for this test.**

Administered *Month Year*

1) What is the product of the square root of 81 and the square root of 49?

    A. 2,05   C. 35

    B. 63   D. 45

2) A bread recipe calls for $2\frac{3}{16}$ cups of flour. If you only have $1\frac{5}{8}$ cups of flour, how much more flour is needed?

    A. $1\frac{1}{2}$ C. $1\frac{1}{8}$

    B. $\frac{9}{16}$ D. $\frac{1}{8}$

3) There are 180 rooms that need to be painted and only 14 painters available. If there are still 12 rooms unpainted by the end of the day, what is the average number of rooms that each painter has painted?

    A. 9   C. 14

    B. 12   D. 16

4) Convert 0.053 to a percent.

    A. 0.5%   C. 5.30%

    B. 0.53%   D. 53%

5) If $4x - 9x + 5x = -29$, then what is the value of x?

    A. Any Number   C. Does not exit

    B. Positive Infinity   D. Negative Infinity

6) Mia is 6 years older than her sister Elise, and Elise is 8 years younger than her brother Mason. If the sum of their ages is 83, how old is Elise?

  A. 23   C. 15

  B. 25   D. 21

7) John is driving to visit his mother, who lives 75 miles away. How long will the drive be, round−trip, if John drives at an average speed of 25 mph?

  A. 30 Minutes   C. 90 Minutes

  B. 180 Minutes   D. 360 Minutes

8) While at work, Emma checks her email once every 80 minutes. In 8−hour, how many times does she check her email?

  A. 3 Times   C. 9 Times

  B. 8 Times   D. 6 Times

9) Julie gives 8 pieces of candy to each of her friends. If Julie gives all her candy away, which amount of candy could have been the amount she distributed?

  A. 259   C. 685

  B. 752   D. 300

10) If a rectangle is 40 feet by 35 feet, what is its area?

  A. 2,350   C. 1,280

  B. 1,400   D. 750

11) You are asked to chart the temperature during a 6-hour period to give the average. These are your results:

6 am: 11 degrees   1 pm: 37 degrees

8 am: 16 degrees   3 pm: 34 degrees

10 am: 28 degrees   5 pm: 30 degrees

What is the average temperature?

A. 37   C. 26

B. 11   D. 24

12) Each year, a cybercafé charges its customers a base rate of $17, with an additional $0.25 per visit for the first 40 visits, and $0.15 for every visit after that. How much does the cybercafé charge a customer for a year in which 60 visits are made?

A. $18   C. $35

B. $15   D. $30

13) If a vehicle is driven 30 miles on Monday, 35 miles on Tuesday, and 25 miles on Wednesday, what is the average number of miles driven each day?

A. 30 Miles   C. 29 Miles

B. 33 Miles   D. 35 Miles

14) What is the prime factorization of 216?

A. $2 \times 2 \times 3 \times 3$ C. 3×5

B. $2 \times 2 \times 2 \times 3 \times 3 \times 3$ D. $2 \times 2 \times 3 \times 5$

15) Three co-workers contributed $14.35, $16.15, and $18.65 respectively to purchase a retirement gift for their boss. What is the maximum amount they can spend on a gift?

    A. 245.05   C. $18.45

    B. $49.15   D. $29.08

16) A family owns 16 dozen of magazines. After donating 48 magazines to the public library, how many magazines are still with the family?

    A. 32   C. 144

    B. 36   D. 768

17) In the deck of cards, there are 4 spades, 7 hearts, 5 clubs, and 8 diamonds. What is the probability that William will pick out a spade?

    A. $\frac{1}{8}$ C. $\frac{1}{9}$

    B. $\frac{1}{6}$ D. $\frac{1}{5}$

18) William is driving a truck that can hold 9 tons maximum. He has a shipment of food weighing 81,000 pounds. How many trips will he need to make to deliver all the food?

    A. 2 Trip   C. 2.5 Trips

    B. 4.5 Trips   D. 3.5 Trips

19) A man goes to a casino with $180. He loses $40 on blackjack, then loses another $90 on roulette. How much money does he have left?

A. $40

B. $50

C. $60

D. $120

20) A woman owns a dog walking business. If 8 workers can walk 16 dogs, how many dogs can 8 workers walk?

A. 8

B. 12

C. 20

D. 16

21) Jude was hired to teach five identical math courses, which entailed being present in the classroom 35 hours altogether. At $30 per class hour, how much did Aria earn for teaching one course?

A. $500

B. $420

C. $620

D. $210

22) If one acre of forest contains 145 pine trees, how many pine trees are contained in 17 acres?

A. 27

B. 2,465

C. 5

D. 135

23) Ava needs $\frac{1}{6}$ of an ounce of salt to make 1 cup of dip for fries. How many cups of dip will she be able to make if she has 32 ounces of salt?

A. 3

B. $\frac{1}{3}$

C. 192

D. $\frac{1}{192}$

24) Two out of 32 students had to go to summer school. What is the ratio of students who did not have to go to summer school expressed, in its lowest terms?

A. $\frac{15}{16}$

C. $1\frac{4}{7}$

B. $\frac{1}{8}$

D. $1\frac{1}{7}$

25) I've got 64 quarts of milk and my family drinks 4 gallons of milk per week. How many weeks will that last us?

A. 7 Weeks

C. 2.5 Weeks

B. 8 Weeks

D. 4 Weeks

AFOQT Practice Test 1

Mathematics Knowledge

- o **25 questions**

- o **Total time for this section: 22 Minutes**

- o **Calculators are not allowed for this test.**

Administered *Month Year*

1) Which of the following is not equal to 6^2?

   A. the square of 6   C. 6 cubed

   B. 6 squared   D. 6 to the second power

2) What's the reciprocal of $\frac{x^5}{32}$?

   A. $\frac{32}{x^5} - 4$ C. $\frac{32}{x^5} + 4$

   B. $\frac{3}{x^3}$ D. $\left(\frac{2}{x}\right)^5$

3) If a = 6, what is the value of b in this equation? $b = \frac{a^2}{2} + 4$

   A. 8   C. 18

   B. 9   D. 22

4) The fifth root of 32 is:

   A. 5   C. 2

   B. 4   D. 3

5) A circle has a radius of 2 inches. What is its approximate area? ($\pi = 3.14$)

   A. 10.72 square inches   C. 30.4 square inches

   B. 12.56 square inches   D. 25 square inches

6) In the following diagram what is the value of x?

   A. 57

   B. 42

   C. 133

   D. 47

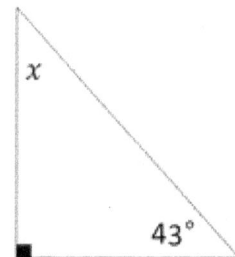

7) If $-8a = 64$, then $a = $ ___

    A. −8   C. 16

    B. 8   D. 4

8) What is 625,460 in scientific notation?

    A. 62.546   C. 0.062546×106

    B. 6.2546×105 D. 0.62546

9) In the following right triangle, what is the value of x rounded to the nearest hundredth?

    A. 10.

    B. 7

    C. 5

    D. 2.40

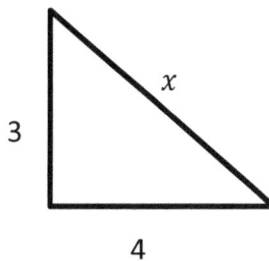

10) Which of the following sets of factors do both 16 and 42 have in common?

    A. {0, 1, 2, 4}   C. {0, 4, 8}

    B. {1, 2}   D. {3, 9, 12, 18}

11) Which of the following is a composite number?

    A. 19   C. 39

    B. 29   D. 59

12) $(2x + 4)(3x + 2) = $?

    A. $5x^2 + 6$ C. $6x^2 + 16x + 8$

    B. $6x^2 + 18x + 8$ D. $7x^2 + 8$

13) $5(a - 4) = 12$, what is the value of a?

   A. 5.14   C. 3

   B. 10.4   D. 6.4

14) If $8^{12} = 2^{15} \times 2^{7x}$, what is the value of x?

   A. 4   C. 5

   B. 1   D. 3

15) The volume of this box is:

   A. 448 cm^3

   B. 252 cm^3

   C. 315 cm^3

   D. 176 cm^3

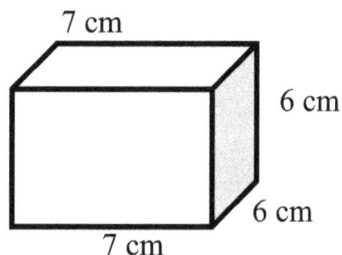

16) Find the slope of the line running through the points (8, 12) and (6, 6).

   A. $-\dfrac{1}{3}$ C. 3

   B. -3 D. $\dfrac{1}{3}$

17) In the following diagram, the straight line is divided by one angled line at 55°. What is the value of $\angle a$.

   A. 55°

   B. 25°

   C. 125°

   D. 315°

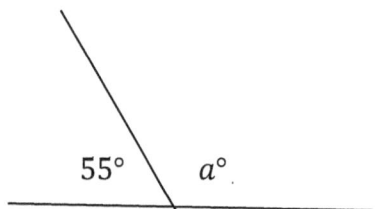

18) Factor this expression: $x^2 + 4x - 5$

A. $x^2(3x - 5)$ C. $(x + 5)(x - 1)$

B. $3x(x + 5)$ D. $(x + 5)(x - 5)$

19) What's the area of the non-shaded part in the following figure?

A. 175

B. 124

C. 64

D. 40

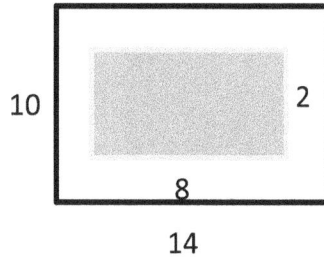

20) A medium pizza has a diameter of 16 inches. What is its area?

A. 12π C. 8π

B. 18π D. 64π

21) What is the circumference of a circle with center at point A if the distance from point X to Y is 12 feet? ($\pi = 3.14$)

A. 72.13 Feet

B. 37.68 Feet

C. 36.26 Feet

D. 34.83 Feet

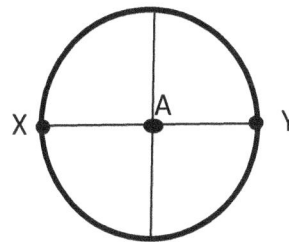

22) Solve for the value of y in the following system of equations.

$$8x + 3y = 5, -6x - y = 5$$

A. -7 C. -2

B. 7   D. 2

23) What is the value of $\sqrt{18} \times \sqrt{48}$?

   A. $40\sqrt{6}$ C. $12\sqrt{6}$

   B. $100\sqrt{14}$ D. $10\sqrt{210}$

24) Which of the following is an acute angle?

   A. $90°$ C. $145°$

   B. $180°$ D. $85°$

25) What is the value of $\frac{8!}{6!}$?

   A. 8   C. 56

   B. 2   D. 180

AFOQT Practice Test 2

Arithmetic Reasoning

- o **25 questions**

- o **Total time for this section: 29 Minutes**

- o **Calculators are not allowed for this test.**

Administered *Month Year*

1) Camille uses a 40% off coupon when buying a sweater that costs $80. If she also pays 8% sales tax on the purchase, how much does she pay?

A. $48

C. $47.50

B. $48.36

D. $51.84

2) Will has been working on a report for 3 hours each day, 5 days a week for 3 weeks. How many minutes has Will worked on his report?

A. 45

C. 2,320

B. 2,700

D. 5,040

3) James is driving to visit his mother, who lives 240 miles away. How long will the drive be, round–trip, if James drives at an average speed of 50 mph?

A. 335 minutes

C. 541 minutes

B. 576 minutes

D. 816 minutes

4) In a classroom of 60 students, 32 are female. What percentage of the class is male?

A. 34%

C. 30%

B. 35%

D. 65%

5) Which of the following is NOT a factor of 36?

A. 18

C. 9

B. 12

D. 14

6) You are asked to chart the temperature during a 6-hour period to give the average. These are your results:

2 am: 7 degrees   12 am: 27 degrees

5 am: 12 degrees   10 am: 19 degrees

6 am: 32 degrees   11 pm: 11degrees

What is the average temperature?

A. 16   C. 18

B. 22   D. 28

7) During the last week of track training, Emma achieves the following times in seconds: 62, 54, 42, 67, 51, and 69. Her three best times this week (least times) are averaged for her final score on the course. What is her final score?

A. 49 seconds   C. 68 seconds

B. 54 seconds   D. 51 seconds

8) How many square feet of tile is needed for a 15 feet x 15 feet room?

A. 225 square feet   C. 150 square feet

B. 140.5 square feet   D. 120 square feet

9) With what number must 1.605687 be multiplied in order to obtain the number 160,568.7.?

A. 100   C. 10,000

B. 1,000   D. 100,000

10) Emma is working in a hospital supply room and makes $40.00 an hour. The union negotiates a new contract giving each employee a 5% cost of living raise. What is Emma's new hourly rate?

A. $49 an hour

C. $38 an hour

B. $45 an hour

D. $42 an hour

11) Emily and Lucas have taken the same number of photos on their school trip. Emily has taken 5 times as many photos as Mia. Lucas has taken 40 more photos than Mia. How many photos has Mia taken?

A. 10

C. 12

B. 25

D. 15

12) Which answer is equivalent to five to the fourth power?

A. 0.0006

C. 0.625

B. 62,500

D. 625

13) Find the average of the following numbers: 15, 26, 22, 37

A. 23

C. 25

B. 26

D. 23.3

14) A mobile classroom is a rectangular block that is 42 feet by 34 feet in length and width respectively. If a student walks around the block once, how many yards does the student cover?

A. 1,600 yards

C. 152 yards

B. 150 yards

D.159yards

15) What is the distance in miles of a trip that takes 4.1 hours at an average speed of 19.5 miles per hour? (Round your answer to a whole number)

    A. 79 miles   C. 50 miles

    B. 80 miles   D. 24 miles

16) The sum of 5 numbers is greater than 100 and less than 160. Which of the following could be the average (arithmetic mean) of the numbers?

    A. 20   C. 33

    B. 23   D. 43

17) A barista averages making 18 coffees per hour. At this rate, how many hours will it take until she's made 1,620 coffees?

    A. 75 hours   C. 90 hours

    B. 80 hours   D. 107 hours

18) Nicole was making $7.50 per hour and got a raise to $7.84 per hour. What percentage increase was Nicole's raise?

    A. 3%   C. 4.53%

    B. 1.57%   D. 3.16%

19) An architect's floor plan uses $\frac{1}{5}$ inch to represent one mile. What is the actual distance represented by $2\frac{1}{2}$ inches?

    A. 12.5 miles   C. 5 miles

    B. 9 miles   D. 4 miles

20) A snack machine accepts only quarters. Candy bars cost 50¢, a package of peanuts costs 70¢, and a can of cola costs 30¢. How many quarters are needed to buy two Candy bars, one package of peanuts, and one can of cola?

A. 9 quarters

C. 8 quarters

B. 6 quarters

D. 4 quarters

21) A writer finishes 180 pages of his manuscript in 60 hours. How many pages is his average per hour?

A. 3

C. 4

B. 8

D. 5

22) I've got 34 quarts of milk and my family drinks 2 gallons of milk per week. How many weeks will that last us?

A. 2 weeks

C. 4.25 weeks

B. 5.25 weeks

D. 4.05 weeks

23) A floppy disk shows 637,045 bytes free and 539,052 bytes used. If you delete a file of size 825,139 bytes and create a new file of size 688,786 bytes, how many free bytes will the floppy disk have?

A. 687,179

C. 884,867

B. 773,398

D. 989,209

24) If $2y + 3y + 8y = -91$, then what is the value of y?

A. −5

C. −7

B. −4

D. −6

25) The hour hand of a watch rotates 30 degrees every hour. How many complete rotations does the hour hand make in 4 days?

A. 8

C. 16

B. 4

D. 18

AFOQT Practice Test 2

Mathematics Knowledge

- o **25 questions**

- o **Total time for this section: 22 Minutes**

- o **Calculators are not allowed for this test.**

Administered *Month Year*

1) Simplify: $6(3x^7)^3$.

    A. $162x^9$ C. $162x^{18}$

    B. $162x^{21}$ D. $162x^{20}$

2) What is the perimeter of the triangle in the provided diagram?

    A. 1,625

    B. 50

    C. 51

    D. 45

Triangle with sides labeled 15, 15, and 15.

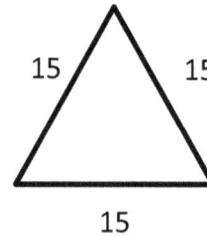

3) If x is a positive integer divisible by 5, and $x < 50$, what is the greatest possible value of x?

    A. 25   C. 45

    B. 15   D. 18

4) $(x + 7)(x + 3) =$?

    A. $x^2 + 10x + 14$ C. $x^2 + 21x + 10$

    B. $2x + 10x + 28$ D. $x^2 + 10x + 21$

5) Convert 670,000 to scientific notation.

    A. 6.7×10^5 C. 6.70×100

    B. 6.70×10^{-5} D. 6.70×1000

6) Which of the following is an obtuse angle?

    A. 19∘   C. 150∘

    B. 85∘   D. 260∘

7) $8^4 \times 8^8 =?$

    A. 8^4 C. 8^{16}

    B. 8^{32} D. 8^{12}

8) What is 2,356.56245 rounded to the nearest tenth?

    A. 2,356.482   C. 2,356

    B. 2,356.6   D. 2,356.562

9) The cube root of 512 is?

    A. 8   C. 888

    B. 88   D. 82,207,728

10) A circle has a diameter of 16 inches. What is its approximate area? ($\pi = 3.14$)

    A. 100.06   C. 49.00

    B. 200.96   D. 14.56

11) There are two pizza ovens in a restaurant. Oven 1 burns six times as many pizzas as oven 2. If the restaurant had a total of 168 burnt pizzas on Saturday, how many pizzas did oven 2 burn?

    A. 13   C. 24

    B. 15   D. 10

12) Which of the following is the correct calculation for 9!?

    A. $3,024 \times 5!$.   C. $1 \times 2 \times 3 \times 6!$

    B. $36 \times 7!$ D. $9 \times 8 \times 8!$

13) The equation of a line is given as: $y = 6x - 3$. Which of the following points does not lie on the line?

A. (1, 3)   C. (3, 15)

B. (−3, −22)   D. (2, 9)

14) How long is the line segment shown on the number line below?

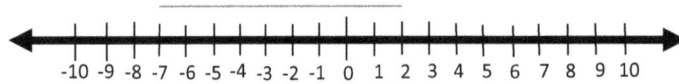

A. −11   C. 2

B. −9   D. 9

15) What is the distance between the points (8, 14) and (3, 2)?

A. 9   C. 12

B. 5   D. 13

16) $x^2 - 81 = 0$, x could be:

A. 9   C. 8

B. 7   D. 13

17) A rectangular plot of land is measured to be 130 feet by 170 feet. Its total area is:

A. 22,100 square feet   C. 38,500 square feet

B.13,404 square feet   D. 34,040 square feet

18) Which of the following is NOT a factor of 164?

    A. 82   C. 41

    B. 4   D. 9

19) The sum of 3 numbers is greater than 130 and less than 200. Which of the following could be the average (arithmetic mean) of the numbers?

    A. 85   C. 45

    B. 70   D. 99

20) One fourth the cube of 8 is:

    A. 56   C. 128

    B. 28   D. 8

21) What is the sum of the prime numbers in the following list of numbers?

      26, 11, 5, 18, 22, 23, 36, 42

    A. 35   C. 43

    B. 27   D. 39

22) Convert 25% to a fraction.

    A. $\frac{7}{20}$ C. $\frac{3}{20}$

    B. $\frac{1}{4}$ D. $\frac{3}{4}$

23) With what number must 4.36598 be multiplied in order to obtain the number 43,659.8?

    A. 100   C. 10,000

    B. 1,000   D. 100,000

24) The supplement angle of a 55∘ angle is:

A. 115°

C. 90°

B. 125°

D. 35°

25) 50% of 40 is:

A. 15

C. 20

B. 45

D. 40

Chapter 12 : Answers and Explanations

Answer Key

In this section, answers and explanations are provided for two AFOQT Math Tests.

Review the answers and explanations to learn more about solving questions fast.

AFOQT Math Practice Test

Arithmetic Reasoning -1				Mathematic Knowledge -1				Arithmetic Reasoning -2				Mathematic Knowledge -2			
1	B	16	C	1	C	16	C	1	D	16	B	1	B	16	A
2	B	17	B	2	D	17	C	2	B	17	C	2	D	17	A
3	B	18	B	3	D	18	C	3	B	18	C	3	C	18	D
4	C	19	B	4	C	19	B	4	B	19	A	4	D	19	C
5	A	20	D	5	B	20	D	5	D	20	C	5	A	20	C
6	A	21	D	6	D	21	B	6	C	21	A	6	C	21	D
7	D	22	B	7	A	22	B	7	A	22	C	7	D	22	B
8	D	23	C	8	B	23	C	8	A	23	B	8	B	23	C
9	B	24	A	9	C	24	D	9	D	24	C	9	A	24	B
10	B	25	D	10	B	25	B	10	D	25	A	10	B	25	C
11	C			11	A			11	A			11	C		
12	D			12	C			12	D			12	A		
13	A			13	D			13	C			13	B		
14	B			14	D			14	C			14	D		
15	B			15	B			15	B			15	D		

Answers and Explanations
AFOQT Practice Test 1
Arithmetic Reasoning

1) Answer: B

$\sqrt{81} \times \sqrt{49} = 9 \times 7 = 63$

2) Answer: B

$2\frac{3}{16} - 1\frac{5}{8} = 2\frac{3}{16} - 1\frac{10}{16} = 1\frac{19}{16} - 1\frac{10}{16} = \frac{9}{16}$

3) Answer: B

$180 - 12 = 168 \Rightarrow \frac{168}{14} = 12.$

4) Answer: C

To convert a decimal to percent, multiply it by 100 and then add percent sign (%).

$0.053 \times 100 = 5.30\%$

5) Answer: A

$4x - 9x + 5x = -29 \Rightarrow 0 \neq -29 \Rightarrow$ The equation not related to x and you can choose any number.

6) Answer: A

Elise = Mia − 6 ⇒ Mia = Elise + 6

Elise = Mason − 8 ⇒ Mason = Elise + 8

Mia + Elise + Mason = 83

Now, replace the ages of Mia and Mason by Elise. Then:

Elise + 6 + Elise + Elise + 8 = 83

3Elise + 14 = 83 ⇒ 3Elise = 83 − 14

3Elise = 69 ⇒ Elise = 23

7) Answer: D

distance=speed ×time ⇒ time $= \frac{distance}{speed} = \frac{150}{25} = 6$

(Round trip means that the distance is 150 miles)

The round trip takes 6 hours. Change hours to minutes, then: $6 \times 60 = 360$

8) Answer: D

Change 8 hours to minutes, then: $8 \times 60 = 480$ minutes

$$\frac{480}{80} = 6$$

9) Answer: B

Since Julie gives 8 pieces of candy to each of her friends, then, then number of pieces of candies must be divisible by 8.

A. $259 \div 8 = 32.375$

B. $752 \div 8 = 94.$

C. $685 \div 8 = 85.625$

D. $300 \div 8 = 37.5$

Only choice B gives a whole number.

10) Answer: B

Area of a rectangle = width × length = $40 \times 35 = 1,400$

11) Answer: C

$$average = \frac{sum}{total}$$

Sum = $11 + 16 + 28 + 37 + 34 + 30 = 156$

Total number of numbers = 6

$$average = \frac{156}{6} = 26$$

12) Answer: D

The base rate is $17.

The fee for the first 40 visits is: $40 \times 0.25 = 10$

The fee for the visits 41 to 60 is: $20 \times 0.15 = 3$

Total charge: $17 + 10 + 3 = 30$

13) Answer: A

$$average = \frac{sum}{total} = \frac{30+35+25}{3} = \frac{90}{3} = 30$$

14) Answer: B

Find the value of each choice:

A. $2 \times 2 \times 3 \times 3 = 36$

B. $2 \times 2 \times 2 \times 3 \times 3 \times 3 = 216$

C. $3 \times 5 = 15$

D. $2 \times 2 \times 3 \times 5 = 60$

15) Answer: B

The amount they have = $\$14.35 + \$16.15 + \$18.65 = 49.15$

16) Answer: C

16 dozen of magazines are 192 magazines: $16 \times 12 = 192$

$192 - 48 = 144$

17) Answer: B

probability $= \dfrac{desired\ outcomes}{possible\ outcomes} = \dfrac{4}{4+7+5+8} = \dfrac{4}{24} = \dfrac{1}{6}$.

18) Answer: B

1 ton = 2,000 pounds

9 ton = 18,000 pounds

$\dfrac{81,000}{18,000} = 4.5$

William needs to make at least 4 trips to deliver all the food.

19) Answer: B

$180 - 40 - 90 = 50$

20) Answer: D

Each worker can walk 8 dogs: $16 \div 8 = 2$

8 workers can walk 16 dogs. $8 \times 2 = 16$

21) Answer: D

$35 \div 5 = 7$ hours for one course. $7 \times 30 = 210 \Rightarrow \$210$

22) Choice B is correct

Write proportion and solve.

$\dfrac{1}{145} = \dfrac{17}{x} \Rightarrow x = 17 \times 145 = 2,465$.

23) Answer: C

Write a proportion and solve.

$$\frac{\frac{1}{6}}{1} = \frac{32}{x} \qquad x = \frac{32}{\frac{1}{6}} = 192$$

24) Choice A is correct

32 students did not have to go to summer school.

$$32 - 2 = 30$$

$$\frac{30}{32} = \frac{15}{16}$$

25) Choice D is correct

1 quart = 0.25 gallon

64 quarts = 64 × 0.25 = 16 gallons

then: $\frac{16}{4} = 4$ weeks

AFOQT Practice Test 1

Mathematics Knowledge

1) Answer: C

Only choice C is not equal to 6^2

2) Answer: D

The reciprocal of $\frac{x^5}{32}$ is $\frac{32}{x^5} = \left(\frac{2}{x}\right)^5$

3) Answer: D

If a = 6 then:

$b = \frac{a^2}{2} + 4 \Rightarrow \qquad b = \frac{6^2}{2} + 4 = 18 + 4 = 22$

4) Answer: C

$\sqrt[5]{32} = 2$

$(2^5 = 2 \times 2 \times 2 \times 2 \times 2 = 32)$.

5) Answer: B

(r = radius)

Area of a circle $= \pi r^2 = \pi \times (2)^2 = 3.14 \times 4 = 12.56$

6) Answer: D

All angles in a triable add up to 180 degrees.

$90° + 43° = 133°$

$x = 180° - 133° = 47°$

7) Answer: A

$-8a = 64 \Rightarrow a = \frac{64}{-8} = -8$

8) Answer: B

In scientific notation form, numbers are written with one whole number times 10 to the power of a whole number. Number 625,460 has 6 digits. Write the number and after the first digit put the decimal point. Then, multiply the number by 10 to the power of 5 (number of remaining digits). Then:

$625,460 = 6.2546 \times 10^5$

9) Answer: C

Use Pythagorean Theorem: $a^2 + b^2 = c^2$

$(3)^2 + (4)^2 = c^2 \implies 9 + 16 = 25 = C^2 \implies C = \sqrt{25} = 5$

10) Answer: B

Factor of 16: $\{1, 2, 4, 8, 16\}$

Factor of 42: $\{1, 2, 3, 6, 7, 14, 21, 42\}$

Then, factors they have in common is: $\{1, 2\}$

11) Answer: A

39: $\{1, 3, 13, 39\}$.

The rest of choices have the factor of 1 and itself.

12) Answer: C

Use FOIL (first, out, in, last) method.

$(2x + 4)(3x + 2) = 6x^2 + 4x + 12x + 8 = 6x^2 + 16x + 8$

13) Answer: D

$5(a - 4) = 12 \implies 5a - 20 = 12 \implies 5a = 12 + 20 = 32$

$\implies 5a = 32 \implies a = \dfrac{32}{5} = 6.4$

14) Answer: D

Use exponent multiplication rule:

$x^a \times x^b = x^{a+b}, (x^a)^b = x^{ab}$

Then: $8^{12} = (2^3)^{12} = 2^{36} = 2^{15} \times 2^{7x} = 2^{15+7x}$

$36 = 15 + 7x \implies 7x = 36 - 15 = 21 \implies x = 3$

15) Answer: B

Volume = length × width × height

Volume = $6 \times 6 \times 7 \implies$ Volume = 252 cm^3

16) Answer: C

Slope of a line: $\dfrac{y_2 - y_1}{x_2 - x_1} = \dfrac{rise}{run}$

$\dfrac{y_2 - y_1}{x_2 - x_1} = \dfrac{6 - 12}{6 - 8} = \dfrac{-6}{-2} = 3$

17) Answer: C

The straight line is 180 degrees. Then:

$\angle a° = 180° - 55° = 125°$.

18) Answer: C

To factor the expression $x^2 + 4X - 5$, we need to find two numbers whose sum is 4 and their product is -5.

Those numbers are 5 and -1. Then: $x^2 + 4X - 5 = (x + 5)(x - 1)$

19) Answer: B

The area of the non-shaded region is equal to the area of the bigger rectangle subtracted by the area of smaller rectangle.

Area of the bigger rectangle = $14 \times 10 = 140$

Area of the smaller rectangle = $8 \times 2 = 16$

Area of the non-shaded region = $140 - 16 = 124$

20) Answer: D

Diameter D= $2r \Rightarrow 16 = 2r \Rightarrow r = 8$

Area = $\pi r^2 \Rightarrow A = \pi(8)^2 \Rightarrow A = 64\pi$

21) Answer: B

Diameter D= $2r \Rightarrow 12 = 2r \Rightarrow r = 6$

Circumference = $2\pi r \Rightarrow C = 2\pi r \Rightarrow C = 12 \times 3.14 = 37.68$

22) Answer: B

$$\begin{cases} 8x + 3y = 5 \\ (-6x - y = 5) \times 3 \end{cases} \Rightarrow \begin{cases} 8x + 3y = 5 \\ -18x - 3y = 15 \end{cases} \Rightarrow -10x = 20 \Rightarrow x = -2$$

$8x + 3y = 5 \Rightarrow 8(-2) + 3y = 5 \Rightarrow 3y - 16 = 5 \Rightarrow 3y = 21 \Rightarrow y = 7$.

23) Answer: C

$\sqrt{18} = \sqrt{9} \times \sqrt{2} = 3\sqrt{2}$

$\sqrt{48} = \sqrt{16} \times \sqrt{3} = 4\sqrt{3}$

$3\sqrt{2} \times 4\sqrt{3} = 12\sqrt{6}$.

24) Answer: D

An acute angle is an angle of greater than 0 degrees and less than 90 degrees. Only choice a is an obtuse angle.

25) Answer: B

Factorial:

$$n! = 1 \times 2 \times 3 \times \ldots \times n$$

$$n! = n(n-1)(n-2)(n-3)!$$

$$\frac{8!}{6!} = \frac{(8 \times 7) \times 6!}{6!} = 56.$$

Answers and Explanations
AFOQT Practice Test 2
Arithmetic Reasoning

1) Answer: D

$40\% \times 80 = \frac{40}{100} \times 80 = 32$

The coupon has $32 value. Then, the selling price of the sweater is $48 ($80 - 32 = 48$).

Add 8% tax, then: $\frac{8}{100} \times 48 = 3.84$ for tax

then: $48 + 3.84 = 51.84$

2) Answer: B

3 weeks = 15 days. Then:

$15 \times 3 = 45$ hours

$45 \times 60 = 2,700$ minutes

3) Answer: B

$distance = speed \times time \Rightarrow time = \dfrac{distance}{speed} = \dfrac{240 + 240}{50} = 9.6$

(Round trip means that the distance is 480 miles)

The round trip takes 9.6 hours. Change hours to minutes, then:

$$9.6 \times 60 = 576$$

4) Answer: B

$60 - 32 = 28$ male students

$\frac{28}{80} = 0.35$

Change 0.35 to percent $\Rightarrow 0.35 \times 100 = 35\%$

5) Answer: D

The factors of 36 are:

$\{1, 2, 3, 4, 6, 9, 12, 18, 36\}$

14, is not a factor of 36.

6) Answer: C

$$average = \frac{sum}{total},$$

Sum = $7 + 12 + 32 + 27 + 19 + 11 = 108$

Total number of numbers = 6

$$\frac{108}{6} = 18$$

7) Answer: A

Emma's three best times are 42, 54, and 51.

The average of these numbers is:

$$average = \frac{sum}{total},$$

Sum = $42 + 54 + 51 = 147$

Total number of numbers = 3

$$average = \frac{147}{3} = 49.$$

8) Answer: A

The area of a 15 feet x 15 feet room is 225 square feet.

$15 \times 15 = 225$

9) Answer: D

$1.605687 \times 100,000 = 160,568.7$

10) Answer: D

5 percent of 40 is: $40 \times \frac{5}{100} = 2$

Emma's new rate is 42.

$40 + 2 = 42.$

11) Answer: A

Emily = Lucas

Emily = 5 Mia $\quad \Rightarrow$ Lucas = 5 Mia

Lucas = Mia + 40

then: Lucas = Mia + 40 $\quad \Rightarrow$ 5 Mia = Mia + 40

Remove 1 Mia from both sides of the equation. Then: 4Mia = 40 $\Rightarrow$ Mia = 10

12) Answer: D

$5^4 = 5 \times 5 \times 5 \times 5 = 625$

13) Answer: C

Sum $= 15 + 26 + 22 + 37 = 100$

$average = \dfrac{100}{4} = 25$

14) Answer: C

Perimeter of a rectangle $= 2 \times$ length $+ 2 \times$ width $=$

$2 \times 42 + 2 \times 34 = 84 + 68 = 152$

15) Answer: B

$Speed = \dfrac{\text{distance}}{\text{time}}$

$19.5 = \dfrac{distance}{4.1} \Rightarrow distance = 19.5 \times 4.1 = 79.95$

Rounded to a whole number, the answer is 80.

16) Answer: B

Let's review the choices provided and find their sum.

A. $20 \times 5 = 100$

B. $23 \times 5 = 115 \Rightarrow$ is greater than 100 and less than 160

C. $33 \times 5 = 165$

D. $43 \times 5 = 215$

Only choice B gives a number that is greater than 100 and less than 160.

17) Answer: C

$\dfrac{1 \, hour}{18 \, coffees} = \dfrac{x}{1800} \Rightarrow 18 \times x = 1 \times 1,620 \Rightarrow 18x = 1,620 \Rightarrow x = 90$

It takes 90 hours until she's made 1,800 coffees.

18) Answer: C

$percent \ of \ change = \dfrac{change}{original \ number},$

$7.84 - 7.50 = 0.34$

$percent \ of \ change = \dfrac{0.34}{7.50} = 0.0453 \quad \Rightarrow 0.0453 \times 100 = 4.53\%$

19) Answer: A

Write a proportion and solve.

$$\frac{\frac{1}{5}inches}{2.5} = \frac{1\ mile}{x}.$$

Use cross multiplication, then: $\frac{1}{5}x = 2.5 \rightarrow x = 12.5$.

20) Answer: C

Two candy bars costs $100¢$ and a package of peanuts cost $70¢$ and a can of cola costs $30¢$. The total cost is:

$100 + 70 + 30 = 200$, 200 is equal to 8 quarters.

$8 \times 25 = 200$

21) Answer: A

$180 \div 60 = 3$

22) Answer: C

1 quart = 0.25 gallon

34 quarts = $34 \times 0.25 = 8.5$ gallons

then: $\frac{8.5}{2} = 4.25$ weeks

23) Answer: B

The difference of the file added, and the file deleted is:

$825,139 - 688,786 = 136,353$

$63,7045 + 136,353 = 773,398$

24) Answer: C

$$2y + 3y + 8y = -91 \Rightarrow 13y = -91 \Rightarrow y = -\frac{91}{13} \Rightarrow y = -7$$

25) Answer: A

Every day the hour hand of a watch makes 2 complete rotation. Thus, it makes 8 complete rotations in 4 days.

$24 \times 30 = 720 \Rightarrow 720 \div 360 = 2$

$2 \times 4 = 8$

AFOQT Practice Test 2

Mathematics Knowledge

1) Answer: B

$6(3x^7)^3 \Rightarrow 6 \times 3^3 \times x^{21} = 162x^{21}$

2) Answer: D

Perimeter of a triangle = side 1 + side 2 + side 3 = 15 + 15 + 15 = 45

3) Answer: C

From the choices provided, 15, 25 and 45 are divisible by 5. From these numbers, 45 is the biggest.

4) Answer: D

Use FOIL (First, Out, In, Last) method.

$(x + 7)(x + 3) = x^2 + 3x + 7x + 21 = x^2 + 10x + 21$

5) Answer: A

In scientific notation form, numbers are written with one whole number times 10 to the power of a whole number. Number 670,000 has 6digits. Write the number and after the first digit put the decimal point. Then, multiply the number by 10 to the power of 5 (number of remaining digits). Then:

$670,000 = 6.7 \times 10^5$

6) Answer: C

An obtuse angle is an angle of greater than $90°$ and less than $180°$.

7) Answer: D

Use exponent multiplication rule: $x^a . x^b = x^{a + b}$

Then: $8^4 \times 8^8 = 8^{12}$

8) Answer: B

2,356.56245 rounded to the nearest tenth equals 2,356.6

(Because 2,356.56 is closer to 2,356.6 than 2,356.5)

9) Answer: A

$\sqrt[3]{512} = 8$

10) Answer: B

Diameter = 16

then: Radius = 8

Area of a circle = πr^2 $\Rightarrow$ A = 3.14(8)^2 = 200.96

11) Answer: C

Oven 1 = 6 oven 2

If Oven 2 burns 6 then oven 1 burns 2 pizzas.

Oven 1 + oven 2 = 7 burn pizzas.

168 ÷ 7= 24 oven 2

168 – 24 = 144 (6 × 24) oven 1

12) Answer: A

$n! = n(n-1)(n-2)(n-3)(n-4)!$

$9! = 9 \times 8 \times 7 \times 6 \times 5! =$

13) Answer: B

Let's review the choices provided. Put the values of x and y in the equation.

A. (1, 3) $\Rightarrow x = 1 \Rightarrow y = 3$ This is true!

B. (−3, −22) $\Rightarrow x = -3 \Rightarrow y = -21$ This is not true!

C. (3, 15) $\Rightarrow x = 3 \Rightarrow y = 15$ This is true!

D. (2, 9) $\Rightarrow x = 2 \Rightarrow y = 9$ This is true!

14) Answer: D

$2 - (-7) = 9$

15) Answer: D

Use distance formula: $d = \sqrt{(x_1 - x_2)^2 + (y_1 - y_2)^2} = \sqrt{(8-3)^2 + (14-2)^2}$

$\sqrt{25 + 144} = \sqrt{169} = 13$

16) Answer: A

$x^2 - 81 = 0$ $\Rightarrow$ $x^2 = 81$ $\Rightarrow x$ could be 9 or –9.

17) Answer: A

Area of a rectangle = width × length = 130 × 170 = 22,100

18) Answer: D

factor of 164 = {1, 2, 4, 41, 82, 164}

9 is not a factor of 164.

19) Answer: C

Let's review the choices provided.

A. $85 \times 3 = 255$

B. $70 \times 3 = 210$

C. $45 \times 3 = 135$

D. $99 \times 3 = 297$

From choices provided, only 135 is greater than 130 and less than 200.

20) Answer: C

The cube of $8 = 8 \times 8 \times 8 = 512$

$\frac{1}{4} \times 512 = 128$

21) Answer: D

From the list of numbers, 11, 5, and 23 are prime numbers. Their sum is:

$11 + 5 + 23 = 39$

22) Answer: B

$25\% = \frac{25}{100} = \frac{1}{4}$

23) Answer: C

Number 4.36598 should be multiplied by 10,000 in order to obtain the number 43,659.8

$4.36598 \times 10,000 = 43,659.8$

24) Answer: B

Two Angles are supplementary when they add up to 180 degrees.

$125° + 55° = 180°$

25) Answer: C

$\frac{50}{100} \times 40 = 20$

"End"

www.ingramcontent.com/pod-product-compliance
Lightning Source LLC
Chambersburg PA
CBHW080510090426
42734CB00015B/3020